W9-AOB-312

Geographies of Disability

Space, and related issues, such as mobility and accessibility, are profoundly important to disabled people's everyday lives, yet this fact has been given little attention within social policy or by urban planners, architects and social science researchers.

Geographies of Disability examines how geography shapes the experiences of disabled people, exploring the relationship between space and disability; how space, place, and issues such as mobility dictate the experiences of disabled people.

Drawing on a significant range of case studies and historical and contemporary data sources, and including illustrative maps and photos, Gleeson clearly presents the key theories and the concerns at the heart of disabled people and their social movements worldwide. The book is organised into three parts. Part I represents a critical appraisal of theories of disability, space and embodiment and develops a disability model. Part II takes a historical perspective and uses case studies to expose how the transition to capitalism affected the everyday lives of disabled people. Part III explores contemporary scenarios of disability: the Western city and the important policy realms of community care and accessibility regulation.

Explaining why the production and development of space has disadvantaged disabled people, both in the past and in contemporary societies, *Geographies of Disability* presents an important contribution to the key policy debates on disability in Western societies, and offers new insights for broader contemporary discussions on embodiment and space.

Brendan Gleeson is Research Fellow in the Urban Research Program, The Australian National University, Canberra, Australia.

Geographies of Disability

Brendan Gleeson

London and New York

First published 1999
by Routledge
11 New Fetter Lane, London EC4P 4EE

Simultaneously published in the USA and Canada
by Routledge
29 West 35th Street, New York, NY 10001

© 1999 Brendan Gleeson

The right of Brendan Gleeson to be identified as the Author of this
Work has been asserted by him in accordance with the Copyright,
Designs and Patents Act 1988

Typeset in Galliard by
Ponting–Green Publishing Services, Chesham, Buckinghamshire
Printed and bound in Great Britain by
T.J. International Ltd, Padstow, Cornwall

All rights reserved. No part of this book may be reprinted or
reproduced or utilised in any form or by any electronic,
mechanical. or other means, now known or hereafter
invented, including photocopying and recording, or in any
information storage or retrieval system, without permission in
writing from the publishers.

British Library Cataloguing in Publication Data
A catalogue record for this book is available
from the British Library

Library of Congress Cataloging in Publication Data
Gleeson, Brendan
   Geographies of Disability / Brendan Gleeson.
         p.   cm.
   Includes bibliographical references and index.
      1. Physically handicapped.   2. Physically handicapped–
History.   3. Architecture and the physically handicapped.
4. Spatial behavior.   5. Human geography.   6. Urban geography.
7. City planning.   I. Title.
   HV3011.G59   1999
   305.9'0816–dc21   98–25641   CIP

ISBN 0–415–17908–4 (hbk)
ISBN 0–415–17909–2 (pbk)

This is for Michael

Contents

viii   *Contents*

Figures and tables

Engineering

Acknowledgements

My friend Gerard Toal described his recent book *Critical Geopolitics* as 'an intellectual product of a scattered journey through various educational institutions' in a range of countries. I can find no more apt description for the origin of this book, whose genesis and writing was something of a geographical odyssey. Various parts of the book were conceived and written in Melbourne, Bristol, Dunedin, Berlin and Canberra, and I owe a debt of thanks to the many communities of colleagues in these scattered places who inspired the intellectual journey that is recorded in *Geographies of Disability*. Among all of these people, it is to Ruth Fincher of the University of Melbourne that I feel the deepest gratitude. In many ways, this book is a testament to Ruth's own outstanding scholarship, and, in particular, the wise and caring advice that she provided during my doctoral studies and later. I must also thank Nigel Thrift, Sarah Whatmore and Paul Glennie for their support and encouragement during a very fruitful research sojourn at the University of Bristol in 1991. Mike Oliver and Paul Abberley also assisted me greatly during this visit, and in subsequent communication. Other colleagues were always helpful on the journey, including Robin Law, Stephen Horton and Donna-Rose Harris in Dunedin, Jane Jacobs in Melbourne, and Ian Cook in Bristol (and later Lampeter).

The writing of the manuscript occurred in two exceedingly stimulating environments: the Urban Research Program at the Australian National University and the Australia Centre at the University of Potsdam. The Urban Research Program is a wonderfully inspirational place for a young academic, due largely to the warm leadership of Pat Troy. I am especially grateful to Pat for his encouragement and counsel during the writing of this book, and also for his support for my stay at the University of Potsdam in 1997. At Potsdam, the staff of the Australia Centre are owed my deepest thanks for furnishing me with just the right place for reflective writing.

Nicholas Brown, Penny Hanley, Kay Anderson and Jane Jacobs read various drafts of the chapters in this book and I thank them all for this. I shudder to think of what the book might have been without their critical input which helped to reduce, if not entirely erase, my own deficiencies in thought and writing. Two anonymous referees also provided very helpful comments on a draft version of the manuscript. In addition to those mentioned, I would like to thank the many others who have supported and encouraged me during my various intellectual meanderings in recent years, including my parents and families, Ulrike Gleeson, Michael Webber, the members of the Disability Action Group (Dunedin), Sarah Lloyd, Casey Mein, Nicholas Low, Michael Dear, Jennifer Wolch, Gerard Toal, George McDonnell, Phil Turton, Pat Fensham, Haydie Gooder, Natalie Jamieson, Chris Philo, Ali Memon and Brian Heenan. Last, and most importantly, I thank Ulrike and Julian Gleeson for the inspiration of their love.

Earlier versions of Chapters 8 and 9 appeared in *Progress in Human Geography* (21, 2) and *International Planning Studies* (2, 3) respectively. They are reproduced in part by kind permission of the journal publishers, Arnold and Carfax Publishing Ltd. Figures 4.1 and 4.2 'The Battle Between Carnival and Lent' by Bruegel are reproduced with permission of Kunsthistorisches Museum, Vienna.

1 Introduction

The purpose of this book

This book is about the relationship between space and disability. In particular, the book explores how social and spatial processes can be used to *disable* rather than *enable* people with physical impairments. The topic is important for at least two reasons: first, space, and related concepts such as mobility and accessibility, are profoundly important to the lived experience of disability; second, this fact has been given relatively little attention in the past by most Western social scientists, including those in the spatial disciplines, Urban Planning, Geography and Architecture. In Geography, the long disciplinary silence on this profound dimension of human experience is especially perplexing. According to the United Nations, there are approximately 500 million persons in the world with physical impairments (Campbell and Oliver, 1996). Moreover, at any given time, disability probably affects 10 to 15 per cent of national populations (Golledge, 1993). Disability is, simply put, a vitally important human experience that Geography cannot afford to ignore. A failure to embrace disability as a core concern can only impoverish the discipline, both theoretically and empirically.

While disability has, until recently, been neglected in the main disciplinary fora – journals, books, study groups, conferences, etc. – there has been a small, but important, tradition of geographic work that, since the early 1970s, has focused on the needs and social experiences of disabled people. I refer here especially to the pioneering work of Reg Golledge, and also to the many unpublished research projects on disability, often undertaken by postgraduates in departments spread across Europe, North America and Australasia. If one widens the conceptual lens for a moment, then the important early work of North American geographers, such as Julian Wolpert, Michael Dear, Jennifer Wolch and Martin Taylor, on mental illness and social dependency also counts as part of this genealogy of disability studies within Anglophone Geography. None the less, these few voices were the exception to the long entrenched disciplinary rule that disability was not a valid geographical concern.

Also, there have been broader social consequences of this disciplinary silence, though this fact is rarely acknowledged. Doubtless the recent undermining of progressive-modernist forms of social science has discouraged open declarations on how academic knowledge can improve people's daily lives, especially those lived in the shadows of injustice or prejudice. Indeed, this, and any other social scientific analysis of disability, can begin from the premise that disabled people throughout the world endure social oppression and spatial marginalisation, facts that will be central concerns in this book. As the United Nations puts it, disabled people:

> frequently live in deplorable conditions, owing to the presence of physical and social barriers which prevent their integration and full participation in the community. As a result, millions of disabled people throughout the world are segregated and deprived of virtually all their rights, and lead a wretched, marginal life.

> (cited in Campbell and Oliver, 1996: 169)

Given the extent of need in disadvantaged communities, it may seem strange that academics, and geographers among them, sometimes seem reluctant to explore certain marginal domains of human experience. To an extent, this reticence is a product of recent critiques that have, quite rightly, questioned the authority of academics who in the past have claimed to speak *for* the 'subjects', or even 'objects', of their research. There remains in the social sciences, a vigorous, and by no means resolved, debate on the tendency of research to colonise, appropriate, and generally misconstrue, the experiences of individuals and groups, especially those whose voices are usually unheard in the discourses of power (e.g., Harding, 1992; Roof and Weigman, 1995). However, this reticence is a relatively recent phenomenon in the social sciences, and therefore does not entirely explain Geography's long avoidance of disability issues. To a large extent, this disciplinary silence reflects the exclusion of disabled people and their concerns from the realms of authoritative knowledge

I would argue that the long failure of geographers to engage with disability issues has denied to disabled people a valuable conceptual, professional and practical resource that might have aided them in their relations – very often, *their struggles* – with the various professional and institutional agencies that have shaped their environments, often in oppressive ways. As many geographers themselves have come to realise, space is a social artefact that is shaped by the interplay of structures, institutions and people in real historical settings. The historical production of space is a contested process where the exercise of power largely determines who benefits and who loses from the creation of new places and landscapes. Knowledge about how space is produced, and for whom, is, of course, a vital element in this constant power struggle. That disabled people in Western societies have largely been oppressed by the production of space is due in part to their exclusion from the discourses and practices that shape the physical layout of societies. Geography, as Imrie notes (1996a), is one such spatial discourse of power that has marginalised disabled people.

Thus, I begin this book by recording my own hope in the emancipatory potential for new spatial studies of disability – what I term here *geographies of disability*.

I argue that new geographic work on disability needs to do more than simply describe the spatial patterns of disadvantage – it must contribute in a variety of ways to a broader political campaign that disabled people, and advocates, are waging in various struggles against the construction of oppressive environments. As Chouinard has put it, there is a need for new spatial research on disability that

> not only unsettles ableist [i.e., oppressive] explanations of social processes and outcomes, but also considers how such knowledge can be used to further political struggles against environments that exclude and marginalize disabled people.
>
> (1997: 380)

To eliminate oppressive spatial practices and knowledges, it is first necessary to explain how and why they occur. Accordingly, the geographies that constitute this volume will seek to explain why the production of space has disadvantaged disabled people, both in the past and in contemporary societies. From this understanding, one can envisage a broad political-theoretical project that would both resist the sources of spatial oppression and articulate new ways of creating inclusionary landscapes and places. This book will not contribute *directly* to that broader political-theoretical process, as this is properly the task of social movements rather than academic observers. I will, however, speculate in the book's conclusion on the sorts of shifts in theoretical and practical research agendas that are necessary if Geography is to contribute to that broader social movement. Thus, I hope that the historical and contemporary studies here will play some indirect role in the larger emancipatory struggles of disabled people. As Harvey observes, 'A renewed capacity to reread the production of historical-geographical difference is a crucial preliminary step towards emancipating the possibilities for future place construction' (1996: 326). Accordingly, the aims of this book are:

- to theorise the broad historical-geographical relationships that have conditioned the social experience of disabled people in Western societies; and
- to describe and explain the social experiences of disabled people in specific historical-geographical settings.

Glimpses of disability

A contribution to theory and politics

In writing this book, I acknowledge, and welcome, the fact that I am participating, if indirectly, in the process that I am seeking to explain: the historical and contemporary production of spaces that have shaped the lives of disabled people. I offer my studies of this process as matters for debate, contributions to a newly politicised production of spaces and places for disabled people and not as canonical statements of how things have been or are now. At this point I think it necessary to record a few remarks on two profound and inescapable limitations on my contributions.

First, I could not, and do not, hope to produce an exhaustive explanation of the relationship between disability and space. The case against this sort of fixed, totalising theoretical account has been well made by other geographers (e.g., Harvey, 1996) and I do not intend to rehearse it here. Simply put, such static, global explanations are not possible, and nor are they politically desirable. Instead, I offer in this book a partial account of the disability–space relation, within a specific set of social and spatio-temporal frames; namely, geographies that focus on the experience of physically disabled people in historical and contemporary Western societies. (This theoretical and empirical specification is elaborated below.) I *do* propose here a broad theorisation of how space informs the experience of disability, but this is a self-consciously open and flexible schema that can only be improved through subsequent critical debate, both within and outside academia. I *do* offer detailed studies of certain historical and contemporary spaces of disability, but these, again, are crafted as explorations whose findings will be sharpened, and perhaps in some instances refuted, by subsequent empirical work.

Given the inevitably partial nature of my theoretical and empirical studies, I acknowledge that the work in this volume offers not much more than a set of glimpses of the range of geographical experiences that shape disabled people's lives. My own view is that a broader, though never complete, appreciation of specific social experiences can only be achieved through a vigorous, reflexive and inter-disciplinary enquiry. While, as I have noted, Geography has been absent without leave from the broad intellectual campaign that in recent decades has sought to explain disability experiences, there are now very encouraging signs within the discipline that things are changing.

Within English-speaking Geography, a small, but growing, community of geographers are arguing that disability must be a critical disciplinary concern (e.g., Butler, 1994; Chouinard, 1997; Cook, 1991; Dorn, 1994; Dyck, 1995; Golledge, 1990, 1991, 1993; Hall, 1994; Imrie, 1996a; Imrie and Wells, 1993a, 1993b; Kitchin, 1998; Parr, 1997a; Vujakovic and Matthews, 1994). A stream of recent disability related sessions at major national Geography conferences attests to the emergent interest in this topic among a stratum of younger geographers.[1] The growing, if still relatively minor, attention given to disability is further confirmed by a recent major text (Imrie, 1996a) and a range of articles in the main learned journals (e.g., Chouinard, 1994; Golledge, 1993; Imrie, 1996b; Imrie and Wells, 1993a, 1993b; Vujakovic and Matthews, 1994). Another milestone was a special issue of the journal *Environment and Planning D: Society and Space* in 1997 that was devoted to the topic of disability.[2] Perhaps the most significant development has been the formation in 1997 of the Disability and Geography International Network (DAGIN) whose main fora have been an electronic mailing list (GEOGABLE) and a web site.[3] By 1998, the GEOGABLE list counted 70 members drawn from a wide range of countries, including the United States, Canada, Britain, Australia, New Zealand, Sweden, and Germany.[4]

The constraints and privileges of authorship

The second source of partiality in this work is the constraint that authorship and identity bring to bear on any individual piece of scholarship (cf. England, 1994). I am not disabled, and, as a white, middle-class male, neither do I directly experience the other major types of social discrimination or disadvantage that bear down upon various oppressed forms of identity. This fact inevitably limits my ability to understand and explain the experience of disablement, in ways that I cannot myself fully appreciate (cf. Drake, 1997). While I have spent many years working with disabled people in a variety of ways (mainly, care services and research), this in no way equates to having lived with a disability. In attempting to 'draw near' the real (i.e., lived) experience of disability, I have over the years practised a strategy – not always successfully – of listening with care and empathy, though never uncritically, to the voices of disabled colleagues and friends. It therefore struck me as good sense when one of the referees who commented upon the written proposal for this book suggested that I consciously weave the spoken and recorded voices of disabled people throughout the analyses. However, upon reflection I could not arrive at a way of consistently doing this that was not somehow rather contrived and gestural, especially in light of the historiographical constraints that shape the different studies. For example, there are very few surviving records of everyday peasant life in feudalism, and I was not able to trace any voices of disabled people for this era. The problem is largely repeated for the industrial capitalist era, though some records survive from this time of disabled people's spoken views. Invariably, these accounts of everyday life are non-autobiographical and their accuracy may be doubted in some instances. I have analysed with some care a few of the more reliable of these recordings in Chapter 6.

I believe that my work here satisfies the referee's suggestion in another way; indeed, through a course that I have always tried to follow in my research on disability. In this book, a specific set of disabled people's voices resonate with authority and, I believe, a good measure of social authenticity. Put simply, it is the voices of disabled theorists which I have invited to speak loudest in the conceptual discourses throughout this book. The conventions of referencing alone will confirm the foregrounding of disabled theorists' voices in my work. It is to be hoped that honour is done in this small way to Paul Abberley, Donna-Rose Mackay, Harlan Hahn and Michael Oliver who have each generously contributed to my intellectual development over many years, in a variety of ways. I also wish here to acknowledge a deeper, not always obvious, debt that I have incurred to the other disabled thinkers who have over the years shaped my approach to this topic, and whose influence is not adequately recorded in the text.[5]

Having enunciated the conceptual constraints that my identity imposes upon me with respect to disability issues, I think it important to acknowledge also the privileges that my position confers, and how these may help to make my contribution a meaningful one. As a highly educated academic, who has benefited from support by relatively well-resourced universities in a variety of countries, I have been able to approach the disability issue with a set of investigative skills and with a relatively privileged level of access to information and other forms of expertise. Of course, these privileges are the product, at least in part, of an inequitable social system which artificially renders education and information as exclusive rather than universal 'goods'. This fact, I believe, imposes a duty upon people such as myself to employ these privileges in the cause of justice; indeed, towards the dismantling of the very systems that unfairly confer social advantages upon a minority. I hope that in these studies, readers, and more particularly disabled people, will find some evidence that I have observed this duty, and through political commitment rather than class guilt.

Inclusions and exclusions

Having declared what I see as the major conceptual constraints on this book, I want now briefly to elaborate the specific social and spatio-temporal boundaries of the studies. Given that one could spend an entire book explaining the specificities that inevitably frame any study of a major social issue, I will not attempt to justify here at great length the exclusions and inclusions that characterise the book. It is to be hoped that the studies themselves will provide these justifications by conveying a sufficient and explicit sense of purpose.

Geographies of which disabilities?

'Disability' is a term which has many different uses in various places and is therefore impossible to define objectively. Disability may refer to a considerable range of human differences – including those defined by age, health, physical and mental abilities, and even economic status – that have been associated with some form of social restriction or material deprivation. This book will adopt a rather focused sense of the term which is often used in the social sciences – here 'disability' refers to the social experiences of people with some form of physical impairment to a limb, organism or mechanism of the body (Oliver, 1990). Thus, the sense of disability used here encompasses impairments that have an organic basis, including those which manifest themselves as physical and intellectual impairments.

This book is primarily about *physical* disabilities. However, I believe that my social geographic explorations of physical disability do have relevance to other disabling experiences. The book will not focus on the question of mental illness, a specific set of health-related conditions and socio-spatial experiences that can be distinguished from physical disability. However, in laying out an initial conceptualisation of disability, the book will briefly review the considerable geographic work that has been undertaken on mental illness, much of which has relevance to the spatial consideration of physical impairment.

Also, the studies will not directly consider the question of chronic illness. I cannot hope to explore all disability experiences here, and neither, of course, would this be appropriate in a single, empirically focused work. A range of commentators, including geographers, have rightly pointed to the heterogeneity of physical conditions and social experiences that are commonly lumped under the 'disability' rubric (Butler and Bowlby, 1997; Dear *et al.*, 1997; Parr, 1997b; Wendell, 1996). These analysts have opposed approaches that avoid or under-state these profound differences. I agree with this criticism to some extent. None the less, there *is* a political need for inclusive theorisations of disability which try to explain the general social forces that bear down upon all 'impaired' bodies. Such broader conceptualisations, as Chouinard (1997) reminds us, can help to forge common, and therefore powerful, political bonds between people with different types of disabilities. However, it is at the empirical level that specifica-tion should occur – here there is need for a sensitivity to the differences of ex-perience that flow from specific forms of impairment and other identity characteristics.

The need to impose sensible and meaningful boundaries on this analysis means that I will focus on long-term, permanent impairments which are not in the first instance health considerations. Indeed, from the 1960s, disabled people and disability groups struggled in many Western countries to separate the issue of disability from questions of health and illness. This has been reflected in the political struggle to shift institutional conceptions of disability from a medical model to a social model. It was both a conceptual and a strategic issue: disabled people were attempting to assert, in the face of a long medicalisation of their experience, that impairment was not an illness, not an issue that should be pre-sided over by the medical establishment. Of course, this avoided – for under-standable strategic reasons – the difficult issue of whether illnesses could be the subject of disabling social relations.

Now, after broad acceptance of the social model and significant political and institutional gains by the disabled people's movement, it seems appropriate to 'destabilise' – in Davis's (1995) terms – the category of impairment that has informed the social model. With the gradual, if often uneven, withdrawal of medical institutions from disability services and debates, a conceptual-political space has been opened for the re-consideration of how health and disability re-late. There are now an increasing number of theorists and activists, especially feminist writers, who insist that 'impairment' should embrace a variety of health-related conditions and experiences, such as chronic illness, though not on terms dictated by medical discourses and establishments (e.g., Butler and Bowlby, 1997; Dyck, 1995; French, 1993a; Morris, 1991; Wendell, 1996). As one key oppo-nent (Oliver, 1996) of the medical model has acknowledged, there is now both the opportunity and the need for a new, and inclusive, model of impairment that can embrace a broad range of disabling experiences. While such a broad, and inevitably contentious, project is beyond the reach of this book, the theorisation of disability presented here is deliberately constructed on flexible and inclusive terms, and I hope that this might contribute in some way to the task of reconceptualising impairment.

As Davis (1995) maintains, the category of impairment is inherently unstable and open to redefinition and expansion. One can keep adding specific conditions and experiences until the category embraces the entire population. Indeed, this can be a politically powerful exercise, to point out to 'non-disabled' people that impairment, broadly defined, is an experience that virtually all of us will have during our lifetimes. None the less, to give conceptual, and therefore empirical, focus to the present investigation, my aim will be to show how permanent physical impairments have been socially constructed in different times and places. Moreover, in examining physical disability I will give less empirical emphasis to intellectual and sensory impairments. There are important conceptual and historiographical reasons for this: principally, the fact that the historical experience of sensory, intellectual and physical impairments has varied in important ways. In making these choices and in excluding conditions such as mental illness and chronic illness from direct consideration in these studies, I am simply declaring a further partiality in the work. It will properly be the task of a broad political-theoretical movement to arrive at a larger conceptualisation of disability that can address the socially constructed disadvantages faced by physically impaired, chronically ill and mentally ill people.

Spatio-temporal focus

There are several important spatial and temporal boundaries to the studies in this book. First, I restrict my analyses to the experiences of disabled people in Western societies, meaning the historical and contemporary social formations of Europe and its major colonies in North America and Australasia. My reasons for doing this are entirely practical – these are simply the societies with which I am most familiar in personal and professional terms. I hope that others will in time produce geographical studies of disability in non-Western contexts.

Also, my focus is primarily, though not exclusively, urban, a choice which again reflects the contexts with which I am most familiar and indeed those in which I am most interested. (Chapter 5 is the only non-urban case study.) While the urban geographic experience of disability has been poorly studied, the situation in rural areas has been given even less attention. None the less, some promising explorations of rural geographies of disability have been made in recent years (e.g., Gant and Smith, 1984, 1988, 1991; Gething, 1997; Wibberly, 1978). It is not too much to expect that in the years to come new rural studies of disability will emerge to challenge and extend the findings made by urban geographic research on this issue.

Finally, my exploration of geographic literature is confined to that produced in English-speaking countries as my familiarity with non-Anglophone Geography is limited. The collection of readings edited by Korda and Neumann (1997) provide one interesting example of non-Anglophone geographies of disability.

Theoretical-political inclinations

As will become obvious in the next few chapters, my approach to spatial analysis is based upon historical-geographical materialism (cf. Harvey, 1996). As a materialist, I believe that the basic historical and geographical organisation of cultural-material life shapes all social experiences, including disability. My reasons for preferring this theoretical frame will become obvious through the critiques of alternative outlooks that are made in the following chapters. Suffice it here to list two grounds in support of this theoretical position. First, this approach extends and complements the materialist theories of disability for which I have the greatest theoretical and political regard. Second, historical-geographical materialism stresses the importance of studying the spatio-temporal basis for any observed social relation. For disability, the theoretical and political advantages of this outlook are apparent, in that materialism rejects naturalistic (e.g., medicalised) and positivistic explanations in favour of empirically grounded, historical-geographical analyses.

A note on terminology

Many readers may wonder why I refer to 'disabled people', rather than 'people with disabilities'. The latter form is a now common terminological practice in most Western countries, and has been adopted in many official and institutional settings. Its supporters claim that this mode of expression is preferable to 'disabled person' because it emphasises the individual's 'personhood' over the fact of disability. The practice is thus said to be a humanising one that supports the general quest for cultural respect and equal rights by disabled people.

I do not choose to contradict this practice lightly, as I support its general aspiration for the cultural revalorisation of disabled people. I also lament the offensive and exclusionary ways in which disabled people have been referred to in the past. However, several disabled commentators have persuaded me not to adopt this form of address, at least for the time being. Both Abberley (1991a, 1991b) and Morris (1993a), for example, question the validity of such 'rhetorical humanism' when disabled people, by reason of oppression, have their humanity denied to them in virtually every social, economic and cultural arena. These writers suggest that use of the term 'disabled people' serves a political purpose by foregrounding the oppression – in other words, the socially imposed disability – that bears down upon impaired people. As Morris (1993a: x) puts it, the term 'disabled people' has political power because it places 'emphasis on how society oppresses people with a whole range of impairments'. I will briefly revisit this issue in the next chapter.

Plan of the book

This book is organised in three parts. The first presents a socio-spatial model of disability, based upon a critical appraisal of social scientific theories of disability, space and embodiment. This theoretical framework guides the empirical studies in Parts II and III of the book which examine how impaired bodies were/are socialised in specific historical and contemporary settings. The case studies in Part II focus on the historical experiences of disabled people in feudal England and the industrial city. The final part of the book explores three contemporary scenarios of disability: the Western city and the policy domains of community care and access regulation.

Chapter 2 is a critical examination of social scientific explanations of disability. The discussion explores how disability has been conceptualised, historically and geographically. The chapter will first review the various definitions of disablement that have been forwarded in recent decades, charting the shift from individualised, pathological accounts to idealist and social constructionist explanations. From this, a critical review of the field of disability studies will be made. Next the discussion will consider the tradition of geographic thought on disablement, charting the sporadic past interest of the discipline in disability which has lately developed into a serious, and rapidly growing, area of enquiry.

The third chapter outlines a historical-geographical account of disablement. Importantly, this framework is not a transhistorical, totalising account of disability. Rather, the historical-geographical approach is a method of enquiry that encourages a critical and contextualised examination of how individuals, communities and institutions negotiate the conditioning influence of structures and thereby produce unique social spaces. In this, the social evaluation of bodily differences, such as impairment, is seen as crucial to the production of distinctive spaces of experience (places, communities, etc.).

Chapter 4 provides a conceptual introduction to the second part of the book which deals with the historical experience of disability in feudalism and industrial capitalism. In this chapter, I distil from the framework developed in the previous part of the book a set of historiographical principles which can guide the study of disability in past societies. To achieve this, I first critically evaluate the contemporary historiography of disability. From this critique, I then present an outline of my alternative historical-geographical method of analysis.

The fifth chapter presents a historical-geographical analysis of the social space of disability in feudal England. The empirical frame for this investigation is the everyday experience of disabled peasants in rural England of the middle ages. The argument is that while disabled people shared the burdens of exploitation and immiseration suffered by much of the feudal peasantry, they were not structurally oppressed by reason of their physical impairment. As is shown, feudal social space was a relatively porous structure which permitted cultural and economic contributions from people with a great range of bodily capacities, including those with disabilities. Evidence for this claim is drawn from a range of primary and secondary sources, including the Poor Law surveys of Norwich (1570) and Salisbury (1635), both of which reveal the presence of disabled people who remained *in situ* (i.e., within affective networks) and earned income.

Chapter 6 examines the social space of disability in the industrial capitalist city. It is first argued that the rise of commodity relations progressively – sometimes violently – dissolved feudal social space, and thereby lessened the ability of disabled people to make meaningful contributions to their families and households. The chapter then explores the experience of disabled people within the proletarian social space of the industrial city, focusing upon the specific (colonial) case of nineteenth-century Melbourne. As in Chapter 5, this analysis explores the relationship between the general social space of the subaltern orders and the quotidian realm of disabled people. Substantial primary data sources are consulted, drawn from records of everyday life within the homes, workplaces and institutions of colonial Melbourne.

Chapter 7 begins the third part of the book and shifts the focus to contemporary Western societies. The aim of this chapter is to present *an* urban geography of disablement; that is to say, a potential framework for understanding the oppressive experiences of disabled people in contemporary and recent Western cities. It is argued that disablement – the oppressive experiences of physically impaired people – is deeply inscribed in the discursive, institutional and material dimensions of cities. These realms of oppression include an inaccessible built environment, landscapes of dependency (i.e., the frameworks of social support provided by state, private and voluntary bodies), exclusionary modes of consumption and production, and devalorising cultural imagery and public policies. The discussion will emphasise how disabled people, and their allies, have countered these forms of oppression through their own urban social movements that have focused on key policy issues, such as civil rights, accessibility, and open employment. The chapter concludes by offering an alternative vision of produced space based upon a political-ethical ideal I term *enabling justice*. From this, I outline in broad terms the features of *enabling environments* – non-oppressive, and inclusive social spaces.

Chapter 8 applies historical-geographical analysis to a contemporary policy domain, community care. The discussion will critically analyse the claim that community care reduces the social injustice experienced by disabled people. This claim can be challenged through a socio-spatial analysis of policy practice in a range of Western countries. As I show, community care policies can help provide enabling environments for disabled people, but empirical analysis shows that this potential is being limited in most Western countries by a variety of reactionary social-political forces.

In Chapter 9 I further examine the urban context of disablement through a case study of accessibility regulation. As with the previous case studies, this empirically based investigation will demonstrate the power of historical-geographical analysis to explain how specific dimensions of disablement arise from, and are reproduced through, the interplay of structural, institutional and contextual conditions. In this case, the spheres of accessibility regulation (access laws, building standards, rights-based guarantees) are examined, with a view to explaining the origins of inaccessibility in capitalist cities. I also show how this form of oppression is both reproduced and challenged through institutional and political practices. The experience of accessibility regulation in one contemporary New Zealand city, Dunedin, provides an empirical focus for the study. I extend the Dunedin case study through further theoretical and policy analysis, pointing to important parallels and divergences in other policy settings, notably the United Kingdom, the United States and Australia.

The final chapter has two aims: first, to recapitulate the theoretical and empirical arguments of the book; and second, to consider the ways in which the discipline of Geography might play an enabling role in disabled people's struggles for social justice and respect. Of course, it would be presumptuous of me to be too prescriptive in this – the recent upsurge of interest in disability among geographers has already demonstrated the discipline's potential to contribute meaningfully to the lives of disabled people. In the fields of Cultural and Social Geography, my colleagues are rapidly formulating a set of enabling research agendas and theoretical debates. What most interests and concerns me are those remaining sub-disciplinary areas that are yet to engage the question of disability seriously. It is these residual 'silent spaces' that I wish to address in closing the book, with the hope that eventually no geographer will be able to claim that disability is irrelevant to their work.

Part I

A socio-spatial model of disability

2 Social science and disability

Introduction

This chapter will explore how disability has been theorised in the social sciences, notably disability studies (embracing sociological perspectives), History and the spatial disciplines (Geography, Urban Planning and Architecture). Obviously I limit myself by confining this analysis to a chapter-length survey, and the review is necessarily selective. Economics and Anthropology, for example, are not dealt with in any detail, though individual contributions from both disciplines are examined at various points in this chapter and in the remainder of the book. Both disciplines have shown very little theoretical interest in the question of disability.

The aim of this critical review is to identify the elements of a geographically and historically informed social model of disability. Thus my survey of social scientific accounts of disability will mainly focus on theoretical analyses, rather than on the policy-oriented discussions that tend to dominate discourses on impairment in Western societies. With this chapter I take the first major theoretical step in Part I towards the construction of a historical-geographical materialist account of disability. This 'embodied materialism' will be outlined in the following chapter.

The chapter consists of two main parts. The first provides a critical review of the field of disability studies. However, the rather unbounded character of disability studies makes it a difficult theoretical terrain to appraise. Therefore this review will outline some of the field's major theoretical contours by mapping a cross section of significant (i.e., widely cited) contributions from a variety of social scientific commentators. The implied basis for this critical review is the historical-materialist account of disablement which has been developed in recent years by British disability scholars, such as Michael Oliver, Paul Abberley, Len Barton and Vic Finkelstein. After critically evaluating disability studies, I will outline this distinctive historical-materialist perspective.

After the review of disability studies, the discussion will shift to the spatial sciences, focusing on Geography, and to a lesser extent, Urban Planning and Architecture. I will substantiate the previous chapter's criticism that the spatial sciences have largely ignored disability, and also trace the emergent geographies of disability which are promising to correct this long disciplinary silence.

Disability studies

Disability studies is a relatively recent phenomenon, emerging as a 'coherent'[1] discourse in the 1950s (though studies of disability, especially in Anthropology, were known previously – e.g., see the studies by Evans-Pritchard (1937) and Hanks and Hanks (1948)). The rise of the civil rights movement in the United States during the 1960s did much to encourage the growth of a discernible field of disability studies. However, disability studies remains in the United States mostly a discourse on policy issues, such as employment, physical access, benefit rights and deinstitutionalisation.[2]

Disability studies is a cross-disciplinary endeavour[3] with the major points of contact limited to journals, professional networks and conferences. The lack of disciplinary boundaries is a potential advantage, allowing disability studies the freedom to integrate the rather arbitrary divisions of thought institutionalised in Western academies (e.g., between Political 'Science' and Economics).

Theoretical development

Disability studies is a form of enquiry which has long drifted in atheoretical currents (Barnes, 1995; Davis, 1995; Radford, 1994). This is in part due to the fact that many of its contributors are either practitioners (e.g., social workers) or advocates. Both groups of observers tend to focus on the immediate policy land-scape. In recent years several serious considerations of the epistemological dimensions of disability have been made (e.g., Bickenbach, 1993; Davis, 1995, 1997; Rioux and Bach, 1994). Many of these recent contributions to the social theorisation of disability have been by disabled academics (e.g., Abberley, 1991a, 1991b, 1993; Hahn, 1989; Oliver, 1990, 1992, 1996; Shakespeare, 1994; Zola, 1993). However, the broad field of disability studies remains dominated by discussions of policy matters, often conducted within discursive circles of disability professionals (e.g., Dalley, 1991; Smith and Smith, 1991).

The failure of the social sciences generally to consider physical impairment as an important issue partly explains the atheoretical cast of disability studies. This may be seen as part of the wider problem of the entrenched indifference of social science to issues of human embodiment (Frank, 1990; Turner, 1984, 1991).

Some important consequences of the theoretical unconsciousness of disability studies have included long neglect of critical social dynamics, including gender, class and race. However, this situation began to change slowly from the 1970s, and then more rapidly during the 1980s, when a series of empirically grounded analyses by disability commentators focused on mainstream social scientific concerns – including gender (e.g., Campling, 1981; Deegan and Brooks, 1985; Wendell, 1989), age (e.g., Walker, 1980), race (Thorpe and Toikka, 1980), education (e.g., Anderson, 1979) and class (e.g., Townsend, 1979). Although primarily cast within a policy framework, these investigations of critical socio-cultural aspects of disablement laid the empirical and conceptual groundwork for a sociological approach to disability. The sociological turn, which gathered strength in the 1980s, represented an important departure from a tradition of disability commentary that had drawn heavily upon variants of methodological individualism (e.g., psychopathology) (Leonard, 1984; Oliver, 1990).

Thus, a movement towards consideration of other social identities – and the multiple subjectivity of disabled people – has been gathering momentum in recent years.[4] This has doubtless been inspired by the political experiences of practitioners, advocates, and, more importantly, disabled people themselves. The growing visibility in Western countries of social movements based upon coalitions of the marginalised has no doubt encouraged an increasingly broad view of oppression among disability commentators (cf. Abberley, 1991a; Barnes, 1996; Young, 1990).

Hahn (e.g., 1989) has made some particularly thoughtful surveys of the common political ground which might potentially link, if not unite, minority social movements. Abberley has also emphasised the link between disability and other forms of social identity, remarking that, 'This abnormality is something we share with women, black, elderly, gay and lesbian people, in fact the majority of the population' (1991a: 15).

Feminists have made perhaps the most powerful theoretical and empirical exploration of how disability intersects with other identity forms (e.g., Boylan, 1991; Cass *et al.*, 1988; Cooper, 1990; Deegan and Brooks, 1985; Fine and Asch, 1988; Hillyer, 1993; Lonsdale, 1990; Meekosha, 1989; Orr, 1984; Williams and Thorpe, 1992). In confronting the 'double handicap' of gender and disability, these analyses have challenged both the masculinist nature of disability studies and feminisms that have failed to recognise disability as an identity form. The work of Morris (e.g., 1989, 1991, 1992, 1993a, 1993b, 1996) is especially notable for its insistent and critical engagement of masculinist representations of disability. Outside social science, disabled women's experiences and representations have been recorded in a variety of forms, including commentary and biography (e.g., Finger, 1991; Mairs, 1995).

Recently, feminist and other disability perspectives have been powerfully challenged by black disabled people in Britain (e.g., Stuart, 1992, 1993). These new accounts have argued that 'their experience of disability can only be understood within the context of racism' (Oliver, 1996: 142). Gay and lesbian people have recently levelled similar criticisms against disability studies (Hearn, 1991).

While it is clear that disability studies still remain in a state of theoretical underdevelopment, it must be pointed out that the dominance of the field by policy concerns represents both a weakness *and a strength*. The latter quality should never be underestimated. The theoretical discourses which have emerged within disability studies have been firmly rooted in the world of everyday social practice. Though often expressed in theoretically unsophisticated terms, or without reference to major debates in the social sciences and humanities, the analyses of many disability scholars are frequently marked by a *first-hand* grasp of the social context of their enquiries.

Thus, by its nature, disability studies strongly challenges the social theorist by demanding explanations that lead to policy prescription and material change. The highly politicised (if often at a somewhat timorous policy level) nature of disability studies holds great potential for a more theoretically informed praxis. A powerful force for this politicisation has been the increasing numbers of disabled people making influential contributions to the field from critical theoretical perspectives (e.g., Abberley, 1987, 1993, 1997; Appleby, 1994; French, 1993a; Hahn, 1986, 1987a, 1987b; Hevey, 1992; Morris, 1991, 1993a, 1993b; Oliver, 1986, 1990, 1996).

The struggle to define disability

One daunting characteristic of disability studies, and also of Western disability policy realms, has been the seemingly endless shifts in definitional orthodoxies concerning the meaning of terms such as 'disability', 'impairment' and 'handicap' (Oliver, 1990). It could be argued that this definitional complexity, not to say confusion, derives in part from the theoretical underdevelopment of disability studies. For example, earlier recourse to wider social scientific debates on nature–culture relations might have helped disability commentators to refine their conceptualisations. However, it is important to acknowledge why disabled people have placed so much emphasis on the definition issue. Disabled people have objected to, and contested, official constructions of their subjectivities by institutions, such as social service providers, because these understandings have often been innaccurate and offensive. As with other marginalised groups, such as gays and lesbians, disabled people have long endured oppressive constructions of their identity within institutional settings and through codifications in law. It is not surprising, then, that disability politics and disability studies have been marked by an acute awareness of the political importance of definitions, reflecting their role in a broader socio-cultural construction of identity (Oliver, 1990).

Thus one could write an entire book about how disability, and its various synonyms, have been defined and deployed, both theoretically and in policy practice.[5] It is not my intention here to attempt an exhaustive survey of definitional debates – readers can consult a range of already published discussions which have focused on this issue (e.g., Oliver, 1990; Wendell, 1996). Rather, my aim here is to draw attention to one profound shift in definitions of disability, the decline in support among theorists, activists and policy makers for individualised, medicalised accounts in favour of various social models.

There have been many critiques in recent years levelled against the medical explanations of disability that informed law and institutional practice in Western countries until recent decades. As Abberley explains, the medical model 'locates the source of disability in the individual's supposed deficiency and her or his personal incapacities when compared to "normal" people' (1997: 1). In broader social scientific terms, the medical approach was allied with methodological individualism, a long entrenched conservative form of social explanation. Although falling out of official favour in recent years, the influence of medical models is still evident in important official definitions, such as that promoted by the World Health Organisation (WHO). While the WHO model shifts the conceptual focus to functional 'disabilities' and social 'handicaps', the primary causal emphasis is still on physical 'impairments' as the source of an individual's everyday limitations. In contrast to the quasi-medicalised 'functional' approach, the social model 'focuses on the fact that so-called "normal" human activities are structured by the general social and economic environment, which is constructed by and in the interests of non-impaired people' (Abberley, 1997: 1).

This profound change in the social scientific explanation of disability has followed broader theoretical shifts on gender, sexuality and race in recent decades. Broadly, there has been a movement away from theories that have explained social differences as reflections of nature, in particular, the varied bodily characteristics of humans. Many variants of social constructionism have replaced these naturalistic explanations. However, while there is little contemporary support for the idea that social difference is a straightforward product of physical difference, there is broad acceptance in social science for the proposition that embodiment is linked to distinct subjectivities and experiences.

Not surprisingly, the social model has found favour among disability activists and advocacy organisations, such as the Disabled People's International (Barnes, 1996). For disabled people, the social model is inherently politicising and valorising, insisting that disability is a real social identity – i.e., a subjectivity – rather than an objective fact of nature which must be endured, or at best ameliorated. The new sociological approaches to disability of recent decades have been encouraged by the social model's rise in concert with an increasing political dynamism in the broad and diverse disability movements that have emerged in Western countries (Campbell and Oliver, 1996).

However, as will be shown, there are many possible social *models* of disability. Several questions must be addressed in formulating a social model that can actually explain everyday reality. What precisely constitutes the 'general social and economic environment' that Abberley speaks of? And how do these environmental factors 'structure' notions of human normality? A variety of models can emerge from the process of answering these, and other, key questions. The answers given reflect broader debates in the social sciences and in the humanities about social explanation.

Obviously it would be impractical here to try to provide an exhaustive typology and description of the possible social models of disability. But several obvious frameworks or explanatory tendencies can be identified and my intention is to do this as a means of specifying what I perceive, from a historical-materialist perspective, to be some of the pitfalls and dead-ends of the social model.

A critical review of four social models

The structuralist view

First, there is the danger of reducing the entire experience of disability to macro social phenomena, such as the economy, culture, policy systems or institutional practices. This 'structuralist fallacy' obviously reflects a broader tendency that has been rightly criticised in the social sciences and humanities for its inaccurate and dehumanising portrait of people as simple products of social forces. Interestingly, from the perspective of disability, the structuralist fallacy relies on a *disembodied* form of explanation which denies that the human form plays a role in shaping social experience. Even social models which advance under the banner of materialism, emphasising the importance of concrete practices in everyday life, can overlook the body – human corporeality – as a critically important substance and signifier in social processes.

Bickenbach warns that, in some hands, the social model 'oversteers and detaches disablement from its biomedical foundations' (1993: 14). Without supporting structuralist models, Oliver (1996) does offer some explanation for the tendency, arguing that the strong emphasis on social causation has served a strategic function; namely, undermining the authority of medical construc-tions, and the notion that disability is an 'illness' which can be healed, or at least ameliorated, through health technologies and practices. If 'oversteering' has occurred at times, it is doubtless attributable to the struggle of disabled people, against a powerful medical establishment, to make the point that 'Disability as a long term social state is not treatable medically and is certainly not curable' (Oliver, 1996: 36).

None the less, in shifting the emphasis of explanation from a naturalised con-ception of human deficiencies to the everyday construction of social life, we must not abandon the body and neglect the critical fact that it plays a foundational, if historically and spatially specific, role in the constitution of human society. Each body provides a unique set of pathological capabilities and limitations that informs the social experience of the individual – the point is that geographical and historical differences mean that these *corpo*realities correspond to unique *social* realities, i.e., distinct embodiments at different points in time and space. In the next chapter, I will explain more fully how an embodied materialism informs a specific social model of disability.

Humanisms

Another potential dead-end offered by the social model is the sort of humanism that has thrived in policy realms and activist networks in recent years. This ap-proach is revealed in the regular announcements that currently favoured collec-tive and individual terms for disabled people are in need of immediate replacement by 'less dehumanising' alternatives. Typical of this is the insistence by many commentators on terms which primordially stress the humanity of disabled people – e.g., '*people with disabilities*'.

Abberley (1991a, 1991b) has provided a thoroughgoing appraisal of this vari-ant of humanism, and rejects the now popular notion that the term 'people with disabilities' is politically and ethically superior to the term 'disabled people' (the same may be said for the singular form). Abberley (1991a, 1991b) argues that this 'humanisation' of terminology effectively depoliticises the social discrimina-tion to which disabled people are subjected. He is not prepared to accept the displacement of the adjective 'disabled' until disabled people are actually permitted to experience social life in fully human ways. Again, in this approach, the shift away from medical explanations has involved eschewing the importance of the body as a form of material difference in favour of a disembodied humanism which pleads for the equal treatment of social unequals.

Idealism

Another broad social model of disability emphasises the non-material dynamics (e.g., attitudes, aesthetics) that supposedly characterise the human experience of impairment. This work has been sourced in idealism, a philosophy which presumes the human environment to be the product of ideas and attitudes (Gleeson, 1995a). Hevey declaims against idealist explanations of disability where 'the material world (for disabled people, the material world of physical inaccessibility) is taken as given and fixed and is an artefact of the world of attitudes and ideas' (1992: 14).

Social psychology, for example, has inspired a formidable idealism in disability studies. For commentators who subscribe to a social psychology view, disability is viewed as an ideological construct rooted in the negative attitudes of society towards impaired bodies (Abberley, 1991a, 1993; Fine and Asch, 1988). While 'social forces' are acknowledged as constitutive dynamics, their material contents are overlooked in favour of psychological or discursive structures (Meyerson, 1988). The most notorious example of social psychology is the explanation of disability advanced by the interactionist perspective, whose chief evangelist was Goffman (e.g., 1964, 1969).

For Goffman, an individual's 'personality' is said to arise from social interaction – as an iterative process between actors – where attitudes are formed on the basis of the perceived attributes (positive and negative) of others (Jary and Jary, 1991). In this view, disability is understood as a 'stigma' – a negative social attribute or sign – that emerges from the ritualistic interaction of actors in society. Thus, interactionists, like Goffman, were able to posit the reality of a '*disabled personality* moulded by an infinity of stigmatising encounters' (Abberley, 1991a: 11) (emphasis added). Abberley rightly dismisses this view for its idealism, evidenced both by its inability to offer any satisfactory explanation of belief formation (interactionism merely describes this), and by its failure to appreciate the materiality of social practices (such as 'interaction').

The interactionist fallacy of explaining disability as the product of aesthetic and perceptional dynamics has found wide favour in disability studies. Warren exemplifies this tendency with his remark that 'handicap should not be "objectified", not be made a "thing out there in the world", but rather be seen as a matter of interpretation' (1980: 80). Similarly, Deegan and Brooks (1985: 5) suggest that the social restrictions of disability are enforced by 'a handicapped symbolic and mythic world'.

Idealism has profound political implications. The view of disability as an attitudinal structure and/or aesthetic construct avoids the issue of how these ideological realities are formed. Idealist prescriptions are consequently reduced either to the ineffectual realm of 'attitude changing' policies or the oppressive suggestion that disabled people should conform to aesthetic and behavioural 'norms' in order to qualify for social approbation.

This last point invites consideration of a further social model that has had enormous influence in disability studies. At issue is the service principle of 'normalisation', more latterly known among some of its adherents as 'social role valorisation' (Wolfensberger, 1983, 1995).

Normalisation

The final social model that I want to examine here derives from the principle of social role valorisation. This rather verbose sounding model began life with the revealing epithet, 'normalisation', and was described by Wolfensberger and Thomas (1983: 23) as 'the use of culturally valued means in order to enable, establish and/or maintain valued social roles for people'. As the original title suggests, this service philosophy – which has been taken up with great vigour in much of the Western world since the 1970s[6] – has the normalisation of socially devalued (or 'devalorised') people as its object.[7] The appeal to 'culturally valued means' to improve the social position of groups such as disabled people effectively forecloses on the possibility of their challenging both the established norms of society and the embedded material conditions which generated them. 'Normality', as the set of 'culturally valued social roles', is both naturalised and reified by this principle.

Abberley (1991a: 15), speaking as a disabled person, admonishes 'normalising' philosophies and service practices for failing to locate 'abnormality ... in the society which fails to meet our needs'. These perspectives assume, instead, that abnormality resides with the disabled subject. Abberley (1991a) insists that humans are characterised by varying sets of needs which cannot be described through references to 'norms'. Hillyer (1993) agrees, criticising the normalisation principle for its indifference to the heterogeneity of embodiments and needs.

Abberley (1991a: 21) argues that disabled people do not desire the current social standard of 'normality', but rather seek a 'fuller participation in social life'. For many disabled people (especially historical materialists like Abberley), the predominant bourgeois mode of social life is neither 'normal', nor is it one to which they aspire (Abberley, 1993). This is to echo Young's (1990) influential critique of normative political theories which have effaced the critical fact of human social difference by presupposing abstract, homogenised notions of human subjectivity.

The disregard for history

One general characteristic of disability studies – shared by the various social models just reviewed – is the disregard for history (Scheer and Groce, 1988). The ahistorical nature of disability studies can be attributed, at least in part, to the fact that most disability scholars have tended to focus upon applied and policy-oriented research to the exclusion of social theory. As Abberley (1987: 5) has remarked, 'Another aspect of "good sociology" … generally absent is any significant recognition of the historical specificity of the experience of disability' (1987: 6). In an earlier article, Abberley was more specific about the historical unconsciousness of disability studies:

> A key defect of most accounts of handicap is their blind disregard for the accretions of history. Insofar as such elements do enter into accounts of handicap, they generally consist of a ragbag of examples from Leviticus via Richard III to Frankenstein, all serving to indicate the supposed perennial, 'natural' character of discrimination against the handicapped. *Such 'histories' serve paradoxically to produce an understanding of handicap which is … an ahistorical one.*
>
> (1985: 9) (emphasis added)

As Abberley is aware, disability studies have not entirely erased history; they have, however, trivialised the past to the point where it is little more than a reification of the present. Generally, however, two broad types of historiography are evident within disability studies: 'microscopic' histories and historical materialist accounts. The first strategy is by far the more common and is characterised by the type of apriorism and speculation that Abberley (1985) refers to. The usual form is for a commentator to present a few paragraphs on the 'history of disability' (usually restricted to Western societies, though the ambitious are not usually so restrained) by way of prefatory remark to a more contemporaneous study. There are many examples of the 'microscopic history' approach (e.g., Harrison, 1987; Laura, 1980; Lonsdale, 1990; Safilios-Rothschild, 1970; Smith and Smith, 1991; Topliss, 1982).

The chief defects of these historical sketches include brevity, lack of empirical substantiation, theoretical underdevelopment, and reification (through idealist tendencies). While there is neither time nor need to explore all of these deficiencies in detail, it is worth pausing to consider certain of the consequences that these studies have had for the historical consciousness of disability enquiry. Importantly, the limited historiography of disability studies has burdened the field with a number of highly questionable orthodoxies about the social context of impairment in previous societies. The most pernicious of these orthodoxies naturalises disabled people's contemporary social marginality and poverty by depicting them as fixed, historical conditions that have been present in most, even all, past human societies. These orthodoxies will be subjected to critical review in Chapter 4.

'The creatures time forgot'[8]

The social sciences – in particular, History – must themselves accept a large measure of responsibility for the indifference to the past in disability studies. This has been recognised by several disability commentators, including Haj (1970), Oliver (1990) and McCagg and Siegelbaum (1989).[9] Haj is notable for his early recognition of the disabled body's absence from the historical discourse: for him, disability represented 'a vast uncharted area ... of ... history' (1970: 13). This observation, it seems, was to go unheard as twenty years later Oliver (1990: xi) felt compelled to claim that 'On the experience of disability, history is largely silent'. It seems that only a few historians (e.g., Garland, 1995; Riley, 1987) have acknowledged that the issue of impairment in past societies has been largely ignored. Garland (1995), invoking Foucault, has described the historical experience of disability as a 'subjugated history'.

The few attempts made at considering the historical dimensions of disability hardly amount to an adequate treatment of the issue. The early study by Watson (1930), while interesting for its empirical content, is both atheoretical and condescending towards its pathologised subject. In it 'the cripple' is portrayed as a transhistorical problem which different cultures have had to deal with ('the cripple' and 'civilisation' are revealingly juxtaposed in the book's title).

Haj's (1970) study of *Disability in Antiquity* is much less patronising towards its subject. Haj carefully circumscribed his interesting study by concentrating on disability in Islamic Antiquity. While Haj's historical and cultural scope is much more limited than Watson's, his analysis is far richer in theoretical terms. However, like Watson's (1930) chronicle, Haj's investigation never seems to have come to the attention of disability studies.

In the past two decades, new historical investigations of disability have begun to emerge, including Edwards's (1997) and Garland's (1995) studies of the Graeco-Roman world, Norden's (1994) survey of modern cinema, Dorn's (1994) socio-spatial chronicle of the development of American capitalism, and explorations of pre-modern Europe by Winzer (1997), Davis (1995) and Nelson and Berens (1997). The collection edited by Mitchell and Snyder (1997) also contains several historical essays on disability, drawn from a range of periods and places. Generally, the emphasis in these new histories is on past cultural representations of disability – previous political-economic constructions have received little attention. In Chapter 4 I will address one early study that has influenced more recent histories – Stone's (1984) 'statist' chronicle of disability policy in Western countries.

Historical materialist approaches

Recognising the failings of the social models detailed above, a range of theorists, mostly in Britain, have proposed various historical materialist explanations of disability. The analyses of Abberley (1985, 1987, 1991a, 1991b, 1997), Barnes (1991), Finkelstein (1980), Hevey (1992), Leonard (1984), Oliver (1986, 1990, 1996), for example, all draw, to varying degrees, upon this analytical framework which was originally developed by Marx and Engels (e.g., 1967). That much of the historical materialist work has been British-sourced probably reflects the long participation by many disability activists in that country with socialist politics (cf. Campbell and Oliver, 1996). In North America, the recent contributions of Dorn (1994) and Davis (1995, 1997) have adopted the historical materialist position to varying degrees.

Materialists argue that disability is a social experience which arises from the specific ways in which society organises its fundamental activities (i.e., work, transport, leisure, education, domestic life). Attitudes, discourses and symbolic representations are, of course, critical to the construction of this experience, but are themselves materialised through the social practices which society undertakes in order to meet its basic needs. According to Oliver, disabled people's social experiences cannot be understood merely through resort to 'personal histories', or even through ideological or symbolic systems, but must rather 'be located in a framework which takes account of their life histories, their material circumstances, and the meaning their disability has for them' (1996: 139).

Of critical importance is the assertion that disability is both a socially and historically relative identity that is *produced* by society:

> The production of disability ... is nothing more nor less than a set of activities specifically geared towards producing a good – the category disability – supported by a range of political actions which create the conditions to allow these productive activities to take place and underpinned by a discourse which gives legitimacy to the whole enterprise.
>
> (Oliver, 1996: 127)

These materialisms avoid the errors of structural reductionism, which were outlined earlier, by highlighting the socialisation of (impaired) embodiment as the key process through which disability is produced.[10] Materialists have developed the following twofold definition of disability that embodies this idea:

> *Impairment*, lacking part of or all of a limb, or having a defective limb, organism or mechanism of the body;
>
> *Disability*, which is the socially imposed state of exclusion or constraint which physically impaired individuals may be forced to endure.[11]
>
> (Oliver, 1990: 11)

From this, disability is defined as a form of oppression which any society *might* produce through the social constitution of its natural bases (including human bodies). Materialists foreground the mode of production – i.e., Classical Antiquity, Feudalism, Capitalism – as an historically evolving ensemble of political-economic and cultural relations that has structured the social understanding and experience of impairment. Importantly, the *social*, rather than merely individual or even institutional, creation of disability means that structural dynamics, such as production and consumption relations and cultural outlooks, are implicated in its constitution and reproduction.

Both Oliver (1990) and Abberley (1997) have distanced themselves from social models which reduce the origins of disablement to purely economic causes. In contrast to such reductionism, the materialist perspective is a richer framework that stresses the significance of a variety of material practices and representations emerging from culture, the economy and the state. Indeed, the materialist disability account is broadly similar to the *cultural materialism* of Raymond Williams (e.g., 1978, 1980; see also Milner, 1993). For example, when Finkelstein and Stuart (1996) point to the 'disabling culture' of contemporary capitalism, they refer to an ensemble of materially evident relations and representations, including political economic systems. Davis (1995) elaborates the cultural materialist view, pointing out how disability is socially produced through two interdependent 'modalities' – 'function' and 'appearance'. Hence, disability is characterised both by political economic marginality (and even exploitation) and by cultural devaluation; a set of oppressive, interlocking conditions.

The historical materialist perspective opens up the possibility of an emancipatory politics fixed on the goal of overcoming the oppression of disability. Materialists point out that impairment has not always been equated with dependency, and that material change may liberate disabled people from contemporary forms of oppression. The point is neatly captured in the following recent statement by Barnes:

> impairment is not something which is peculiar to a small section of the population; it is fundamental to the human experience. On the other hand, disability ... is not. Like racism, sexism, heterosexism and all other forms of social oppression, it is a human creation.
>
> (1996: xii)

Changing attitudes is a necessary but, on its own, insufficient step towards the realisation of a non-disabling society. Finkelstein emphasises this idea in outlining the requirements for a transformative political practice which would counter the oppression of disability:

> The requirements are for changes to society, material changes to the environment, changes in environmental control systems, changes in social roles, and changes in attitudes by people in the community as a whole.
>
> (1980: 33)

It is important to realise that historical materialism is an analytical framework, and like the social model of disability (Oliver, 1996), it is not an empirical social theory. The materialist framework outlines a set of basic epistemological and ontological principles to guide the study of societies – notably the importance of seeing social relations as historically contingent and structurally conditioned. However, there remains the task of building social explanation by applying these principles in empirically informed studies of historically and geographically specific contexts.

To date there has been very little historical empirical analysis of the past experience of impairment. Davis (1995) and Oliver (1990) have both explored how the transition from feudalism affected physically impaired people. Both analyses, while insightful, are limited: Oliver's chapter-length survey is at a very broad level of abstraction and relies wholly on secondary sources, while Davis concentrates his far more extensive analysis on the historical experiences of deaf people. In addition, both Abberley (1985, 1987) and Barnes (1991) have made contributions to the historical understanding of disability. My own unpublished investigation, undertaken some years ago (Gleeson, 1993), attempted to extend the historical materialist perspective through an extensive empirical examination of disability in past societies.

Davis (1995), Oliver (1990) and the other materialists contrast the experience of disablement in feudal and capitalist social formations. Feudal society, for example,

> did not preclude the great majority of disabled people from participating in the production process, and even where they could not participate fully, they were still able to make a contribution. In this era disabled people were regarded as individually unfortunate and not segregated from the rest of society.
>
> (Oliver, 1990: 27)

Under capitalism, they argue, impairment has been socialised as a specific form of *social oppression* – disability – which contrasts with other forms of injustice and exploitation based upon class, gender, race or sexuality. Contemporary disability oppression is frequently referred to as ableism, which Chouinard and Grant (1995: 139) define as 'any social relations, practices and ideas which presume that all people are able-bodied'.[12] I will explore the character of contemporary disability oppression in greater detail in Chapter 7.

Several criticisms might be raised against the historical materialist work on disability that has been produced thus far. First, it has produceed little detailed historical-empirical analysis of disability. Second, this variant of materialism has not grasped the importance of space to the constitution of society and human identity.[13] Geographers, of course, are well aware of the irreducibly spatial character of social relations, including those which produce marginalised and oppressed identities (Sibley, 1996). However, as noted in the preceding chapter, most geographers have long ignored the disabled identity. If historians can be implicated in the ahistorical nature of disability studies, then geographers can surely take a large share of the blame for the failure of the disability commentators to take space seriously.

In the next part, I first survey the rather fleeting theoretical engagements with disability within the spatial sciences before the 1990s. After this, I explore the recent turn to disability in Geography. There is already evidence of a variety of approaches to the socio-spatial theorisation of disability, and interestingly, not all of the new approaches support the social model unequivocally.

The spatial disciplines

In the foregoing chapter it was observed that Anglophonic Human Geography has largely overlooked the question of disability. As I shall shortly explain, the other spatial sciences – i.e., Urban Planning and Architecture – have not done much better.

I noted in the previous chapter that Geography's record on disability has not been completely abysmal. Considerable attention has been given to closely related issues, such as mental illness (e.g., Dear, 1977, 1981; Dear and Taylor, 1982; Kearns, 1990; Parr, 1997b; Smith and Giggs, 1988) and access to state social services (e.g., Pinch, 1985, 1997; Smith, 1981; Wolch, 1980, 1990). In particular, the urban geographic studies of service dependency, which emerged in North America and Britain in the late 1970s, paid occasional attention to physical and intellectual disability issues (Dear and Wolch, 1987; Pinch, 1997). Wolpert's studies (e.g., 1976, 1978, 1980) were among the few to extend the interest in spatial patterns of 'service dependency' to disabled people.

While medical geography frequently touched upon issues of impairment – under the rubric of 'illness' – the emphasis was on the spatial epidemiology of physical conditions, rather than the social experience of disability (see, for e.g., Lovett and Gatrell, 1988; Mayer, 1981). The epidemological approach revealed a medicalised, *asocial* view of impairment which reproduced the idea that disability was a 'personal tragedy' inflicted by nature (Dorn, 1994; Park *et al.*, 1998).

Also, from the 1960s, a few, rather isolated, voices tried to draw attention to disability issues in Geography. Among these, Golledge (e.g., 1990, 1991, 1993, 1997) stands apart for his systematic research interest on disability from a behaviouralist perspective.[14] Other published analyses of disability issues included Hill's (1985) phenomenological investigation of sightlessness in the United States and the work by Gant and Smith (1984, 1988), Kirby *et al.* (1983) and Nutley (1980, 1990) on transport mobility in Britain. As Park *et al.* (1998) explain, these earlier explorations of disability were largely positivistic in orientation and paid little attention to the social context of disability. Both Chouinard (1997) and Imrie (1996a) are right to conclude that Geography itself has been complicit in the marginalisation of disabled people from authoritative realms of knowledge.

In the other spatial sciences, specialist attention has, in the past, focused on accessible design ideals in Architecture (e.g., Lifchez and Winslow, 1979; Lifchez, 1987) and on transport mobility issues in Urban Planning (e.g., Brail *et al.*, 1976; Wibberly, 1978). These analyses shared with disability studies a general aversion to social theory and a heavy emphasis on mainstream policy 'solutions' to problems of disablement. On occasion, this meant lumping disabled people together with other 'special population groups', such as the elderly, for the purpose of policy analysis (e.g., Brail *et al.*, 1976; Gilderbloom and Rosentraub, 1990); an approach sometimes evident in Geography (e.g., Gant and Smith, 1988, 1991; Golledge, 1990). This homogenising tendency erased, or understated, profound differences in needs and socio-spatial experiences between 'population groups' (Imrie, 1996a). Also, there was a tendency by architects and planners to reduce disability to a 'built environment problematic', an issue that I will explore in more detail in Chapters 7 and 9.

Just as with Geography, there has been a rise of interest recently in disability issues within academic architectural and planning circles. Recent published analyses of disability in Urban Planning (e.g., Bennett, 1990; Imrie and Wells, 1993a, 1993b, Tisato, 1997), and Architecture (e.g., Kridler and Stewart 1992a, 1992b, 1992c; Lebovich, 1993; Leccese, 1993) attest to this new focus on disability, though the engagement with social theory is still very limited. Like their predecessors, these analyses remain relatively isolated from each other, rather than integrated as a critical discourse, and tend to address debates within the non-spatial social sciences.

Interestingly, there is a considerable, if largely invisible, tradition of postgraduate research on disability. I refer here to the many isolated investigations of disability undertaken in Anglophonic geography and planning departments (e.g., Cook, 1991; Dodds, 1980; Dorn, 1994; Gleeson, 1993; Hill, 1986; Lawrence, 1993; McTavish, 1992; Perle, 1969).[15] For various reasons these student investigations of disability in the past failed to mature as an explicit academic agenda; perhaps partly because some of these students are disabled themselves and experience barriers to academic development (cf. Chouinard and Grant, 1995). Another reason why this enormous potential failed to develop until recently has been a general unwillingness of the discipline to recognise the scholarly importance of disability.

The recent turn to disability

As stated in the previous chapter, there is increasing evidence of an awakening of interest among geographers in the issue of disability. Since the early 1990s significant advances have been made in the theorisation of disability as a socio-spatial phenomenon (Park *et al.*, 1998). I draw particular attention here to the theoretically insightful work of Butler and Bowlby (1997), Chouinard (e.g., 1994, 1997), Dear *et. al.* (1997), Dorn (e.g., 1994), Dyck (e.g., 1995) and Imrie (e.g., 1996a, 1996b, 1996c), Moss (e.g., 1997), Parr (e.g., 1997a, 1997b) and Radford and Park (e.g., 1993). As mentioned in Chapter 1, these theoretical developments have been associated with the rise of new disciplinary networks that aim to promote disability as a geographic concern.

The rise of new geographies of disability consolidates the long-established work of geographers in related fields – including behavioural patterns, health issues, and welfare provision – while also reflecting broader shifts of interest in the discipline, especially towards questions of embodiment (Chouinard, 1997). Dorn's (1994) ambitious outline for a 'cultural geography of the stigmatized body' critically situates the emergent disability interest within a tradition of sub-disciplinary concerns, including medical, behavioural, structurationist, and welfare geographies. His general conclusion is that, while these various modes of investigation produced valuable understandings of policy contexts and disabled people's experiences of these, they also, at times, reproduced, or at least failed to challenge, oppressive institutional representations of impairments. Indeed, I will argue in Chapter 10 that the loose field of 'institutional geographies' still needs to develop a fuller, and thereby more emancipatory, notion of disability if it is to avoid the charge that Dorn rightly levels against it.

The recent rapid escalation of interest among geographers in issues of embodiment (e.g., Ainley, 1998; Duncan, 1996; Johnson, 1989a, 1989b; Nast and Pile, 1998; Pile, 1996) parallels the wider emergence of new cultural and sociological theories of the body. As with disability, the new geographies of embodiment also build upon immanent disciplinary projects, including geographies of time, culture, and human difference (especially gender, race and sexuality). All of these established sub-disciplinary concerns have emphasised the socio-spatial construction of the body (though often in radically different ways), and have thereby helped to nurture the emergent geographies of disability. Many of the recent disability geographies tend to reflect the convergence of these established concerns with a focus on impaired embodiment. Just a few select examples of the increasingly rich and divergent field of disability geographies will demonstrate this point.

In a lengthy critique of radical geography, especially feminist perspectives, Chouinard and Grant (1995: 143) listed disabled women as among the 'Missing sisters in geography'. An emergent feminist geography of disability has begun to recover these 'lost bodies' from the marginalised realms of human experience, though Chouinard (1997) observes that the project has hardly begun. Moss and Dyck have proposed 'a feminist political economic analysis of environment and body as an addition to the critical frameworks emerging in medical geography' (1996: 737). They stress a broadened appreciation of impairment, examining how chronic illness is constituted as disabled identities 'within the context of the wider social political economy' (1996: 737).[16] This work finds resonance in analyses by Chouinard (e.g., 1997, and also Chouinard and Grant, 1995). Other geographers have drawn upon different geographic traditions – Imrie (1996a) has used urban political economy to examine accessibility regulation; Dorn (1994) has undertaken a cultural geographic analysis of disability; while Park and Radford (1997) have used a broadly Foucauldian perspective in their investigations of the historical geography of institutions for intellectually disabled people.

A recent heated exchange between myself (Gleeson, 1996a), Butler (1994), Imrie (1996c) and Golledge (1993, 1996) illustrated the potential for important theoretical differences over how disability is to be conceived in socio-spatial terms (see also the commentaries on this debate by Park *et al.*, 1998 and Parr, 1997a). Broadly speaking, Imrie, Butler and I took issue with Golledge's (1993) outline for a geography of disability, mainly because it lacked a social constructionist perspective. As part of our critiques, we proposed similar, if not identical, socio-spatial models of disability that emphasised the importance of political economic relations. Golledge's (1996) angry reply rejected our criticisms (especially those made by Imrie and myself) and also indicated that we had caused offence by under-appreciating his first-hand grasp of the practical challenges that face disabled people in everyday life.[17] Difficult issues surfaced here centring on the authenticity of knowledge and the problem of speaking outside one's identity – both Imrie and I are non-disabled, while Golledge is a disabled person who has made an enormous contribution to the development of navigational aids for blind people (see Golledge, 1997; Golledge *et al.*, 1991; Swerdlow, 1995). Of course, disputation over authenticity and representation is widespread in the humanities and social sciences, and these issues are likely to prove very challenging to geographers of disability. I will engage with these matters more directly in the last chapter of this book.

The episode I have just related indicates that deep theoretical fault lines may already be appearing in the emergent geographies of disability. In the next and last section of this chapter I will outline an historical-geographical approach to disability. While the historical-geographical approach opposes a number of potential social models of disability – notably those critically reviewed in the previous part of the chapter – I would argue that this perspective's basis in an *embodied materialism* makes it a relevant framework, even a 'meeting ground', for the variety of disability geographies mentioned above.

Conclusion: from critique to theory

So far in this chapter, two broad evaluations have been made. First, the rather diverse field of disability studies was found to be theoretically underdeveloped and largely ahistorical. The general effect of these deficiencies is to limit greatly our ability to understand both how disability is produced through the socialisation of impairment, and how this process might vary in different times and places. This then makes difficult, if not impossible, the political task of rooting out the contemporary structures of disability oppression. For example, our unwillingness to acknowledge the historicity of disability will lead us to assume that impairment is associated with 'natural' social restrictions, even disadvantages. From this assumption comes the conventional view that disabled people will always to some extent be socially dependent, and the best we can do therefore is to try to relieve their impairments, with medicine and adaptive aids, and lessen their disabilities through social support and through human rights initiatives. Across Western countries, the disabled peoples movements have, for some time now, voiced their fairly unanimous rejection of this 'reformism', arguing instead for a transformative politics that aims to remove the social structures which oppress them (Campbell and Oliver, 1996).

This radical political aspiration lays an important duty at the feet of academia, in particular, disability studies. In order to support transformative ideals, it must first be shown that fundamental change is possible, and the most obvious strategy for doing this is to demonstrate *that change has already happened*. In other words, it is politically vital to prove that disability is a socio-historical construct, an oppressive structure that was built at some point, through some era, over the lives of impaired people, and which can therefore be torn down and replaced by inclusive social relations. This is, as I see it, the essence of the challenge before historical materialism in respect of disability: to demonstrate scientifically the historicity of disability through studies of how impairment has been lived in past societies. The parallel task is to construct theoretically informed analyses of how disability is lived and produced in the range of contemporary societies. Of course, it is a historiographical fact that we cannot undertake an empirical study of 'all history'. Human societies are bounded both historically and spatially, and comparative historical analysis must seek to relate geographical contexts in meaningful ways.

This brings me to the second review undertaken in this chapter, the survey of the spatial disciplines. None of these disciplines has really done much in the past to help elaborate a social theoretical understanding of impairment, though the emergent geographies of disability are rapidly correcting this deficit. None the less, there has been little, if any, historical geographical attention to disability, a situation that diminishes our ability to understand the lived experience of impairment in previous societies. This then robs from disability studies, and in turn disabled people's movements, the arguments needed to sustain a transformative politics.

How can Geography contribute to a deepened historical-geographical appreciation of disability oppression? One answer is to say that the new social geographies mentioned earlier are already demonstrating the centrality of space to the understanding of disability. But something more is needed: first, these emergent geographies in the main are focusing only on contemporary Western societies; and second, not all these analyses subscribe to the social model of disability. I therefore argue that a new historical-geographical framework is needed in order to spatialise the social model of disability. Of course, building such a framework involves important political and theoretical choices: in this respect it is to be hoped that my dispositions have been made clear in the preceding reviews. The materialist social model is, in my opinion, the best starting point for a historical-geographical analysis of disability. While this critical model is, by definition, anything but a universal or conventional account of disability, I submit that is has the potential to embrace a variety of perspectives on impairment, including the feminist and cultural accounts now emerging in Geography.

Of course, Geography is hardly a theoretically homogenous endeavour, and the task of spatialising the materialist social model of disability demands a further set of epistemological choices. Not surprisingly, I argue that the most useful spatial perspective for this task is the historical-geographical materialism advocated by a range of contemporary observers, including Harvey (1996), Soja (1989) and Smith (1984). This is not to confine the geographical input to this perspective, however, and I think it vital to remain mindful of the many other 'radical' impulses that have addressed the historical-geographical construction of identities.

In essence, then, my proposed framework derives from a cross-fertilisation of historical-geographical materialism and the materialist disability perspectives that have emerged in the social sciences and humanities. However, I think there is one more task to be undertaken before attempting this theoretical convergence. I consider it necessary to first deepen the existing materialist appreciation of disability by raising the broad question of nature and its significance for historical materialist theory. I argue that the best way of approaching disability, theoretically and empirically, is through a broad historical materialist framework which foregrounds embodiment as a key dynamic through which human societies are produced. Approaching the explanation of disability from a general (materialist) account of the body has the signal advantage of constantly foregrounding the fact that impairment is only one of a range of overlapping embodiments, including those defined by sex, gender, race and class.

In the next chapter I will outline this framework which I shall refer to as an *embodied historical-geographical materialism*. While this term is appropriately descriptive, it is certainly unlovely prose, and later in the book I will simply refer to the framework as *embodied materialism*.

3 The nature of disability

Introduction

The purpose of this chapter is to provide a historical materialist framework for analysing disability. In concluding the previous chapter I argued that such a framework should be drawn from a broader historical-geographical account of embodiment. Furthermore, this more general account should explain how bodies, as natural bases of human existence, are given social significance within particular societies. Of course, any 'society' has specific historical-geographical boundaries, and the socialisation of embodiment therefore can take vastly different forms in different times and places. This is in accord with the materialist model that was reviewed in the previous chapter where disability is seen as part of a broader process of social embodiment – the ascription of roles and representations to body types that varies in time and space. Moreover, the socialisation of human embodiment is seen as part of a larger process through which societies transform their 'natural bases' – literally, their material elements – into real physical and cultural environments.

In order to arrive at a historical-geographical account of embodiment, it will be necessary first to examine two related debates about the production of human societies. The first of these concerns the production of nature, how the material world is constituted in the historical creation of human societies. Of course, this 'debate' can be defined very broadly to embrace a wide range of allied and contradictory theoretical accounts. My interest here is in how historical materialism has conceived the production of nature, and, within this, the human body. There exists now in the social sciences and humanities an ever-sprawling literature on the human experience of embodiment (Harvey, 1996). However, as Davis (1995) points out, these new explorations of embodiment have eschewed the question of disability, in favour of other 'somatic identities', notably the 'sexed body'. Furthermore, many 'social constructionist' analyses of the body draw heavily on variants of idealism. For these reasons, I shall have only very limited recourse to this literature, preferring to 'excavate' a materialist account of embodiment from established social theories.

The second, related, discussion centres on the production of space, a fundamental quality of nature and human society. Again, the potential field of concern is vast and my own enquiry will focus on materialist theories of social space, meaning principally the work of historical-geographical analysts. My aim in exploring these two 'social production' debates is to identify the elements of a historical-geographical account of embodiment. From this account, I intend to distil a more specific framework for analysing disability.

Thus the *analytical framework* I wish to outline in this chapter is in no way a transhistorical, totalising *theory of disability*. As I will show, the historical-geographical approach is a method of enquiry that demands a critical and contextualised examination of how individuals, communities and institutions negotiate the conditioning influence of socio-cultural structures (themselves historically fluid) and thereby produce unique social spaces. In this, the social valuation and devaluation of body types is seen as crucial to the production of distinctive spaces of experience (places, communities, etc.).

The chapter is organised in five main parts. First, the production of nature is explored with a view to situating embodiment within the general process through which humanity socialises its material worlds. Following this, I will outline the elements of an embodied materialism. The third part surveys the production of space with the aim of showing how the socialisation of nature, including bodies, occurs geographically. Following these enquiries, a general outline is sketched for an embodied *historical-geographical* materialism. From this I then distil the elements of a historical-geographical approach to disability.

The production of nature

Not surprisingly, the historical materialist debate on nature begins with the writings of Marx and Engels. Their writings in the nineteenth century initiated a rich tradition of materialist thought on nature. This tradition is too broad to summarise fully in the present analysis, and, indeed, there is no need to do this. As I will show, select reference to the works of Marx and certain contemporary thinkers can draw out sufficiently the elements of a materialist account of embodiment. Although embodiment has rarely been an explicit concern of materialism, it is possible to 'locate the body' in the materialist debates on nature, beginning with Marx.

Marx on nature

Nature, for Marx, exists independently of human experience, but for humanity 'it attains its qualities and meanings by means of a transformative relationship of human labour' (Bottomore *et al.*, 1983: 351). Thus nature is held at once to be both an objective, external reality and also the environment in which human beings satisfy their needs (Wood, 1981). It is through labour – the production and reproduction of material needs – that nature is transformed and becomes an internal reality of human development (Turner, 1984). Marx used the notion of 'two natures' to explain this historical transformation process (Smith, 1984). First, the social practices of each human community are seen as transforming the basic materials – both physical and biological – received from previous societies (Bottomore *et al.*, 1983). These inherited materials – or 'first nature' – include everything from built and natural environments to physiologies. When these materials are received and remade by a succeeding society they become known as 'second nature'. Consequently, almost all of the 'natural world' has been some-how altered through human intervention, and nature indissolubly connected to human society.

Marx was of the view that 'nature is mediated through society and society through nature' (Smith, 1984: 19), with the labour process providing the means for this metabolism. As the means through which labour is effected, the body constitutes an ontological exemplar for the unity of nature and human society. Marx vigorously insisted that humans and nature are both 'the sensuous product of historical and social processes' (Turner, 1984: 232). Thus humans are never counterposed to a static, external world of natural things; a conception which Marx held to be a great fallacy of bourgeois science. Yet neither are humans, as social beings, simple products of natural phenomena. His great emphasis was to reject the view of humans as determined by 'objective laws' of nature. Marx saw nature, not as a law provider, but rather as a field of human practice. Turner explains the point: 'Human practice involves, as Marx noted, the humanization of nature in which nature is appropriated and forced to serve human needs' (1984: 246). 'Human essence', thus, derives not from an immutable natural law, but from the common project of women and men in transforming nature so as fully to develop their species potential.

Marx's suggestion that nature is an arena of material potentialities and prob-abilities lays the groundwork for an ontology of embodiment. Bodies, in the first instance, can be viewed as potentially infinite configurations of natural capabili-ties and limitations ('first nature'). Through the transformation of nature, hu-man beings attempt to transcend organic restrictions and fulfil natural capacities (the production of 'second nature'). Obviously, bodies, as elements of received (first) nature, are socially constituted in this transformation process.[1]

Marx's investigations of capitalism aimed to show how a specific mode of production altered nature so as to deny for much of humanity their species po-tential. His analyses also indicate the process by which these same social relations *oppressively embodied* large sections of their human subjects. I think it worthwhile now to pause and consider briefly the more explicit reflections on the human body in Marx's work in order to better understand the significance of this 'op-pressed bodies' thesis that emerges from his work.

Marx and the body

Marx made few specific references to the body, and historical materialism has also largely avoided explicit consideration of embodiment. However, Foucault was only partly right in declaring that 'Marxism considered as an historical reality has had a terrible tendency to occlude the question of the body in favour of con-sciousness and ideology' (1980a: 59). I find in the general concepts of Marx's social theory many powerful insights on the question of social embodiment. Marx offers a rich portrayal of the development of capitalism which manifests a profound conceptual sensitivity to the fact of human biological diversity.

Some of the most general of Marx's conspicuous reflections on the body are to be found in the *Grundrisse* (1973). In this he clearly holds to a concept of the body as doubly constituted. In its 'organic' state the body is thought to provide the 'natural conditions of the producer's existence' (1973: 489). Thus, the worker's

> living body, even though he reproduces and develops it, is originally not posited by himself, but appears as the presupposition of his self; his own (bodily) being is a natural presupposition, which he has not posited.
>
> (1973: 490)

In an earlier work, Marx (1977: 145) insisted that the human is 'directly a natural being'. Moreover, 'as a natural, corporeal, sensuous, objective being he is a *suffering*, conditioned and limited creature, like animals and plants' (1977: 145) (original emphasis).

How does *social* embodiment occur? For Marx, the individual body is socialised through a lifetime of encounter between the subject's organic physiology and her/his experiences of production and reproduction. However, for some bodies, Marx seemed willing to betray his own conception of nature as historically and socially mediated. It is clear that, like most Victorians, Marx regarded sex difference as the natural basis for a social division of labour between women and men. For him, woman's biology made her *naturally* suitable for certain types of labour (reproduction, domestic work, light industry) while at the same time unfitting her for work of a physically demanding nature. Heavy labour, being beyond woman's 'natural' capacities, interfered with her reproductive potential and thus *morally* compromised her principal social status as mother. Marx was unambiguous on this, asserting that there are 'branches of industry that are specifically unhealthy for the female body or are objectionable morally for the female sex' (quoted in Vogel, 1983: 71).

Despite an otherwise sophisticated view of nature as historically and socially mediated, Marx's approach to the question of embodiment as gender reflected the biologism that characterised Victorian thought in general.[2] Nature in this instance provides the social world with a set of fixed and inviolate biological realities. While Marx strenuously denied that other forms of social being (e.g., class membership) could be reduced to the status of natural fact,[3] he was unwilling to extend this renunciation to the issue of gender.

Marx made further allusions to the 'natural' body and its social significance. In *The German Ideology*, he and Engels refer to 'the division of labour which develops spontaneously or 'naturally' by virtue of natural predisposition (e.g., physical strength), needs, accidents etc., etc.' (Marx and Engels, 1976: 50). The quote should not be interpreted as evidence of further biologism: here physiology is seen to operate on the division of labour in conjunction with a set of historically evolving forces, such as needs. Moreover, the reference to 'accidents' as a conditioning influence intimates Marx's recognition that physiology itself was prone to social change, especially that deriving from political-economic sources. As will be shown below, his later works are alive with explicit references to the 'corporeal power' that capital deployed against the proletariat.

The other main area in which Marx openly remarked upon the issue of embodiment was in his examinations of the impoverished peoples of capitalist society. For Marx, the suffering 'pauper' – that spectral 'other' which haunted the Victorian bourgeois imagination – was testimony to capitalism's oppression and exploitation of the physically vulnerable. His pauper was a polymorph whose many forms included vagabonds, criminals, and prostitutes – the 'lumpenproletariat' – together with those in the proletariat who had failed in the competition to sell labour powers. For the latter, Marx clearly recognised physical infirmity as a principal cause of unsaleable labour power. The physically vulnerable included children (especially orphans), the elderly, and the 'mutilated'. The last described the 'victims of industry, whose number increases with the growth of dangerous machinery, of mines, chemical works etc' (Marx, 1976: 797). According to Marx, these bodily characteristics left their owners with little chance of engaging in mainstream wage labour. Even within the surplus (partially employed or wholly unemployed) working population such persons were doomed to settle in 'the lowest sediment of … the sphere of pauperism' (Marx, 1976: 797).

At a more general theoretical level, Marx's analysis of capitalist social relations suggests much about the proletarian experience of embodiment. His dramatic depiction (especially in the first volume of *Capital*) of the capitalist 'juggernaut' literally pulverising the bodies of working women and men in the relentless transit of accumulation is an arresting, not to say disturbing, vision of social embodiment. Here capital is pictured as a mechanised Leviathan of productive forces, crippling and mutilating the bodies of workers.[4] The view is admirably captured by Marx's collaborator, Engels, in this *locus classicus*:

> Women made unfit for childbearing, children deformed, men enfeebled, limbs crushed, whole generations wrecked, afflicted with disease and infirmity, purely to fill the purses of the bourgeoisie.
>
> (Engels, 1973: 180)

The industrial division of labour is named as a brutalising force, suppressing 'a whole world of productive drives and inclinations' innate to human beings, ultimately leaving the worker a 'crippled monstrosity' (Marx, 1976: 482). Marx is rarely more direct in his condemnation of capitalism than he is in his charge that it 'mutilates the worker, turning him into a fragment of himself' (Marx, 1976: 481). In *The Poverty of Philosophy*, the mechanised division of labour is said to leave its human subjects 'dismembered' (Marx, 1975: 130). A tendency of the capitalist labour process then is to disembody workers psychologically by separating their conscious experience of work from their physiological needs and capacities.

Yet this violence is hardly confined to the worker's psyche or spiritual being. Marx (1976: 341) depicts the progressive crushing of worker bodies through the relentless extension of the working day to and beyond 'the *physical limits* to labour' (emphasis added). Indeed, Marx goes further, in an attack on capitalist technology in which the machines of industrialism are portrayed as voracious forces twisting, stunting, breaking, even consuming, their human operatives. Against these ravenous automata Marx counterposes an image of physically frail workers. The machine ogre is ultimately frustrated only by a set of natural limits in the form of 'the weak bodies and strong wills of its human assistants' (Marx, 1976: 526). Here Marx alludes to the social importance of a corporeal nature, describing the human body as 'that obstinate yet elastic natural barrier' to surplus value extraction (1976: 527).

This theme of 'weak bodies' remains present throughout Marx's analyses in *Capital*.[5] In an implicit sense, Marx's focus is on the proletarian body as exhausted, exploited and crippled by the progress of capitalist accumulation (Davis, 1995). Scarry identifies the 'body in pain' as a meta-theme throughout Marx's work: 'The pressures of the body are conveyed in Marx's writings not by sensory or sensuous description but by numerical description' (1985: 268). Scarry (1985) points to Marx's quantitative surveys in *Capital* of the living and labouring conditions of the proletariat as calibrations of the pressure placed on workers' bodies by industrial accumulation. Here Marx summons an immense amount of evidence, mostly drawn from government reports, on factors ranging from the occupancy rates of workers' housing to the calorific intakes of the subaltern classes, as testimony to the corporeal brutality of capitalism.[6]

The bloodied encounter between workers' 'weak bodies' and the incorporeal jaws of industrial automata is a fundamental theme of Marx's analysis of capitalism. But his theorisation of the laws of capitalist development suggests much more for a theory of social embodiment. The broad theoretical conclusion to be drawn from Marx's investigations is that a complex, and historically uneven, repression of certain forms of embodiment has been a core dynamic of capitalism's development. As I will show in Chapter 6, Marx's enquiry into the process of value creation lays the basis for a materialist account of how the physical body was economically signified during the development of capitalism.

The dialectics of nature

Many of the theoretical issues relevant to an embodied materialism have been canvassed by recent historical materialists. I refer, in particular, to the critique of orthodox materialism advanced by the Italian Marxist, Timpanaro (1975), and its subsequent elaboration in the Anglo world by theorists such as Collier (1979) and Soper (1979, 1981, 1995). At the centre of the argument is Timpanaro's (1975) claim for the importance of nature and biology in the development of social relations.

As Soper (1979) explains, Marxian debates about nature and culture have been haunted in the past by vulgar materialisms where biological 'facts' are read directly onto social phenomena. (As seen above, one of these 'vulgarities' was evident in the work of Marx and Engels.) Recognising this failing, many Western Marxists have remained wary of biology, seeing it as the 'territory' of determinist analyses (Barker, 1981). It is this suspicion that informed the unfavourable response to Timpanaro's views on nature which were forwarded in the 1970s.

However, Timpanaro's work was positively (though not uncritically) received by certain socialist scholars, including Raymond Williams (1978). Another sympathiser, Soper, outlines the essence of Timpanaro's critique of established Western historical materialism:

> in its zeal to escape the charge of biological reductionism, Marxism has tended to fall prey to an antithetical form of reductionism, which in arguing the dominance of social over natural factors literally spirits the biological out of existence altogether.
>
> (1979: 63)

Soper's (1979) own position at the time can be located within the socialist feminist movement which emerged in the 1970s. (Her recent work (e.g., *What is Nature?*, 1995) remains avowedly materialist.) For socialist feminists, orthodox Marxism had tended to reduce the history of capitalism to a chronicle of class divisions, ignoring the central issues of sex and gender cleavages (Eisenstein, 1979; Vogel, 1983). Thus:

> Marxism has tended to offer an economistic account which reduces individuals to their economic and class relations and therefore fails to elaborate upon the many other determinants that produce individual difference within the sameness of those economic and class relations.
>
> (Soper, 1979: 66)

While Marx's work laid the conceptual groundwork for rejecting biologism, this insight was not extended to social embodiments beyond the brutalised body of the worker.

Thus, Soper embraced Timpanaro's (1975) call for the re-establishment of *ontological* materialism – emphasising the connection between biological and social phenomena. The very foundations of Marxist thought were at issue:

> Marxist materialist theory of the relations between the natural and social and psychological sciences must extend beyond a statement of the existential primacy of the physical and biological, for it is Marxist not simply by virtue of its recognition of the prior determination of the natural and biological ... but by virtue of its capacity to provide knowledge at the level at which these general determinations make their appearance as specific and always 'socially mediated' effects within society.
>
> (Soper, 1979: 71)

How then to elucidate the interactions between these natural and social worlds? As a first step Timpanaro (1975) confirms Marx's general rejection of any reductionist association between the two. Furthermore, as Soper explains,

> he [Timpanaro] is quite prepared to cede that biology is for the most part 'socially mediated' and that our biological constitution is itself subject to evolution even if its 'history' proceeds at a much slower pace than that of the history of society.
>
> (1979: 68)

Thus, the natural world is held at any moment to pre-exist social formations, but is itself historically changing through both internal evolution and human intervention. Recalling Marx's 'weak bodies' theme, Timpanaro (1975) asserts the ontological primacy of certain biological 'givens' – he names illness, old age and death – as forces that inevitably shape social relations. For Timpanaro, each human relates to nature through the fact of heredity 'and, even more, through the innumerable other influences of the natural environment on his (sic)[7] body and hence on his (sic) intellectual, moral and psychological personality' (1975: 41).[8] Timpanaro maintained a dialectical view of the culture–nature relation where environment and biology are always socially mediated, and vice versa. Both biology and society change over time, though Timpanaro (1975) and Soper (1979) insist that these two transformations occur at different temporal scales. Nature, relative to society, is characterised by historical constancy.

For Soper, materialism insists that the body be taken seriously, even if many materialists have ignored this fact. With its deep appreciation of the sensuousness of human experience, materialism challenges the 'puritanism and elitism of the idealist refusal of the body' (1995: 91). Thus, Soper's enduring concern with the body (e.g., 1979, 1995) has focused on how the presocial human form is rendered a cultural 'artefact'. She is interested in how particular physical attributes – such as 'somatic instincts', physiognomy, physical strength and sex – are socially constituted.

While socialist feminists have repudiated Freud's declaration that 'anatomy is destiny', Soper argues that they have equally opposed any Marxist attempt to reduce sexual relations to economic relations. This rejection of gender as either biologically or socially determined is consistent with Soper's dialectical ontology of culture. The notion of a dialectic implies a mutually constitutive relation between nature and culture. Thus, Soper suggests that anatomy could, in a sense, be redefined as 'destiny', 'in the sense that biological sex difference does and will always have its effects upon human society' (1979: 84). Accordingly, there are no direct correspondences between organic and social embodiments; only socially mediated relations exist. Embodiment should be seen as conditioned by a complex set of historically and socially specific natural elements.

What then are the social processes which generate this metamorphosis of nature? This question foregrounds the defining essence of historical materialist enquiry itself. Historical materialism 'asserts the causal primacy of men's and women's mode of production and reproduction of their natural (physical) being in the development of human history' (Bottomore *et al.*, 1983: 324). From this, the importance of the labour process, as the transformative nexus between nature and culture, is immediately apparent. Marx regarded the act of labour as the 'moment' in which the human body struggles to wrest from nature its immanent potential, while also attempting to transcend its limiting qualities. Moreover, in this process the worker's own body, as a set of 'natural forces', realises (or is denied) its human potential; that is to say, the act of labour socially embodies human beings.[9]

> Labour is, first of all, a process between man and nature, a process by which man, through his own actions, mediates, regulates and controls the metabolism between himself and nature. He confronts the materials of nature as a force of nature. *He sets in motion the natural forces which belong to his own body, his arms, legs, head and hands,* in order to appropriate the materials of nature in a form adapted to his own needs. Through this movement he acts upon external nature and changes it, and in this way he simultaneously changes his own nature. *He develops the potentialities slumbering within nature*, and subjects the play of its forces to his own sovereign power.
>
> (Marx, 1976: 283) (emphasis added)

While Soper agrees that the biological body finds social expression through labour, she insists upon an inclusive definition of 'work' that contrasts with the one often employed by Marx and many subsequent materialists. Soper rightly conceives the labour process in broad terms as embracing all the human endeavours (production and reproduction) that sustain a social system. This field of endeavour acts as

> a kind of exemplar for a materialist approach to questions about the way in which one should think of the forms of unity between biological and social determinations that are instantiated both socially and individually.
>
> (Soper, 1979: 78)

Of course, labour is structured by power relations, and its social division entrenches ruling interests. In its different forms, therefore, labour valorises certain identities and oppresses or devalues others. As Soper explains, this social valorisation – the attempt to preserve class and gender hierarchies – is closely linked to processes of cultural embodiment which produce 'higher' and 'lower' bodies (cf. Stallybrass and White, 1986). Thus, in Western history the 'rejected bodies' of 'the "lower orders" of society, necessarily figure as something less than human: as an uncouth, simple peasantry, or proletariat, whose closeness to the earth and its animals also places it nearer to nature' (Soper, 1995: 91). According to Marx 'the exploitative division between "mental" and "manual" labour' in capitalism was both reflected in, and sustained by, the mind-body dualism that characterised bourgeois philosophy and culture (Soper, 1995).[10]

An embodied historical materialism

The foregoing discussion has identified the key considerations for an embodied historical materialism. In this section I will now draw these elements together before going on to consider the question of space and its meaning for an embodied materialism.

First, it is clear that Marx's view of nature, and its transformation in human society, provides a starting point for any analysis of social embodiment. In this the organic body appears as a set of material capacities and limitations, potentially infinite in its various forms, which is expressed as a social being through its participation in (or exclusion from) the human transformation of nature. In this, the body is viewed as an ontological duality – as both biological and social, as neither solely object or subject. As Benthall puts it:

> The body is a kind of philosophical anomaly in nature. It is both *object* – something with a given weight and texture and dimensions, a given horse power and breaking strain – and *subject*, for the body is (in Merleau-Ponty's words) 'my point of view on the world', that through which there are objects. Or – more correctly – it is neither object nor subject.
>
> (Benthall, 1976: 160) (original emphasis)

The same body is conceivable as being at once both a biological fact and a cultural artefact; the former constituting a pre-social, organic base upon which the latter takes form.

Importantly, the body surfaces insistently within materialist analysis as the irreducible 'site' of material practice, including the production and registration of social power relations. Thus, Foucault, the great materialist theorist of embodiment,[11] mused, 'I wonder whether, before one poses the question of ideology, it wouldn't be more materialist to study first the question of the body and the effects of power on it' (1980a: 58). Indeed, Foucault's investigations of social embodiment in different epochs were intended to show that human corporeality is a *historically and socially specific* phenomenon. Gallagher and Laqueur capture this historical materialist postulate admirably:

> Not only has [the body] been perceived, interpreted, and represented differently in different epochs, but it has also been lived differently, brought into being within widely dissimilar material cultures, subjected to various technologies and means of control, and incorporated into different rhythms of production and consumption, pleasure and pain.
>
> (1987: vii)

Foucault demonstrated the *historical relativity* of social embodiment through historical essays dealing with medicine (1975), insanity (1988b), punishment (1979) and sexuality (1980b, 1986, 1988a). In a broad sense, his investigations echoed the 'corporeal power' theme evident in Marx's work. In the modern era, the forces shaping the social constitution of the body were conceptualised as moving at two distinct, but inter-related, levels: first, within a 'micro-politics' of the regulation of the body; and second, through 'bio-powers', or a structural surveillance and control over populations which served both to maintain and enhance institutional power and to facilitate the accumulation of capital. According to Foucault, certain institutional power structures, operating at both social levels, had emerged in late eighteenth-century Europe aimed at producing compliant subjects ('docile bodies') for industrialising societies. 'Bio-power' was partly realised through a distinct form of industrial spatiality, or 'Panopticism', that encouraged subjects to internalise routines of self-surveillance and compliance with authority.

An important feature of the embodied framework is an imperative to explain the historical construction of natural limitations and opportunities (of which the biological body is a part). Importantly, these 'possibilities' cannot themselves be regarded as an ensemble of asocial and supra-historical 'givens', but must be seen rather as part of the evolving relationship between humanity and nature.[12]

The history of material cultures has been marked by an evolution of the body in both its natural and social forms (Illich, 1986). Just as the progress of human social development has involved the transformation of nature, so too has the organic body evolved in form. However, as a *social construct*, the body has been through a far more rapid and heterogeneous process of transformation than it has as a physiological form. For materialism, the course of social embodiment has not been naturally given, but rather the outcome of a dialectical historical relationship between the natural world and human society. An embodied materialism highlights the importance of labour, broadly defined, as the fundamental crucible within which nature is transformed by human agency, and from which the social body emerges.

Accordingly, a further postulate of this analytical framework must be the insistence that individual labour-powers are socially generated. This point is explained by Soper:

> the capacity to labour is, as Marx himself is the first to insist, in every importantly differentiating sense, socially and not genetically determined. Differences in physical strength, patience, endurance and so on between individuals are as nothing compared to the difference in 'capacity' that results from the social organisation of work, and the technologies in which labour power is harnessed.
>
> (1981: 193)

It is through the *social organisation of work*[13] that each human being – as a unique incarnation of capacities and inabilities – is socially calibrated in terms of its capacity for labour; that is to say, its immanent labour-power.

At a more immediate (historically specific) level of abstraction, Marx's theorisations of capitalist development have demonstrated the explanatory power of an embodied materialism. His enquiries depict a mode of production which has historically privileged certain biological forms of embodiment. This outline for an embodied materialism broadly addresses the relationship between bodies, temporality and society. However, a critical dimension of analysis is missing; namely, the question of space, itself a fundamental quality of nature that is produced through human endeavour into culturally specific forms. In the next section I will address the materialist account of the production of space with a view to providing then a fuller, *historical-geographical* account of embodiment.

The production of space

Historical-geographical materialism

As with the body, the importance of space was long under-appreciated in the historical materialist tradition (Harvey, 1996). In the postwar era, however, this situation began to change, both through a rising interest in space among some materialist thinkers, and through the emergence of Marxian Geography from the early 1970s (Soja, 1989). For Harvey, the convergence of these reformist influences has given rise to a new, spatially enriched form of materialism: 'Historical materialism has to be upgraded to historical-geographical materialism. The historical geography of capitalism must be the object of our theorising' (1989a: 6).

As Soja (1989) has shown, the work of the French theorist, Henri Lefebvre (1901–91), provided a powerful source of inspiration for the emergent historical-geographical materialism. The central elements of Lefebvre's theory of space are to be found in *La Production de l'espace*, which appeared in 1974 (the English edition of 1991 is referred to hereafter). During his lifetime Lefebvre struggled against the structuralist Marxism of Poulantzas, Althusser and, later, Castells. As part of this effort he sought to contradict the structuralist assumption that space was nothing more than the 'mere territorial projection of social relations' (Martins, 1982: 163). Against this 'holographic' view Lefebvre proposed space as a dynamic material force which animated (and therefore delimited) social life. The argument was that society and space were mutually constitutive forces: 'Space is permeated with social relations; it is not only supported by social relations, but it also is producing and produced by social relations' (Lefebvre, 1979: 286). From this, Lefebvre asserts that historical societies produce their own spatialities, just as much as they create observably unique forms of material practice. Indeed, it is through each society's unique social practices that materially-different spaces are produced.

The view that space is much more than the passive Newtonian 'receptacle' of social relations – that society and space are engaged dialectically – has been taken up in the work of many Anglophonic geographers (e.g., Harvey, 1989a, 1990, 1996; Massey, 1984; Soja, 1989; Smith, 1984) and some sociologists (e.g., Gottdiener, 1985). Soja has explained the 'socio-spatial dialectic' to Anglophone readers as a concept 'which recognizes spatiality as simultaneously ... a social product (or outcome) and a shaping force (or medium) in social life' (1989: 7).

The historical and social contingency of produced space is obscured by the tendency of societies to perceive evident spatiality as natural, and thus inevitable. In speaking of the historical approach to space in capitalism, Lefebvre had this to say:

> Social space has ... always been a social product, but this was not recognised. Societies thought that they received and transmitted natural space.
>
> (1979: 286)

Materialists share the idea that space is socially produced with a broad and diverse grouping of 'post-positivist' geographers and social scientists (Goodall, 1987). It is important to distinguish, therefore, the dialectical view of social space employed by materialists from those evident in other theoretical viewpoints.

Social space

According to Buttimer (1969), the concept of social space was first articulated by Durkheim in the 1890s. His use of the term was not precise, but was intended to mark a distinction between socially created space and '"real" space, by which he meant physical space' (Smith, 1984: 75). Aside from a limited life within the works of certain French geographers in interwar, and early postwar, years (Buttimer, 1969), the concept of social space was largely dormant until its revival by humanistic and critical geographers in the late 1960s.

Materialist understandings of space are distinguished from purely idealist or nominalist constructions. In the latter, space is understood as social insofar as it can be demonstrated to be perceptional and experiential. Outside Geography, for example, social space has often been portrayed by humanists as an ideational, or even discursive, construct (e.g., Bachelard, 1969; Ross, 1988[14]). Gregory (1981: 16) has noted that 'The materiality of social life is weakly developed in modern humanism.' Smith (1979: 367), in a similar criticism, has remarked that humanistic geography has too frequently displayed an 'inability to convey coherently the brutal objectivity of much everyday experience'.

The tendency of social space to be associated with idealist geographies (particularly phenomenological approaches) has led both Smith (1984) and Soja (1989) to express deep misgivings about use of the term in materialist studies. Smith (1984: 75) muses that 'social space seems to be spatial only in a metaphorical sense'. Soja (1989: 80n), on the other hand, is troubled by the fact that social space is 'murky with multiple and often incompatible meanings', and in its stead he proposes the term 'spatiality'.

For Lefebvre, social space is materialist in that it is nothing less than the *socialisation of nature*. Social space, for him, is a manifestation of (first) nature transformed through human material practice; that is to say, the 'spatio-temporal rhythms of nature transformed by social practice' (Lefebvre, 1991: 117). Lefebvre insists that social space is a material product that is also dynamically creative – that is, it is both produced socially and socially producing.

At a more concrete level the question may be asked: how is this material force produced? Precisely through the primary processes that transform first nature in general: 'It is the forces of production and the relations of production that produce social space' (Lefebvre, 1991: 210).

Thus Lefebvre is able to define social space as 'the space of social practice, the space of the social relations of production and of work and non-work' (1991: 225). Obviously then, social space is historically relative in that different modes of production must produce unique spatialities.

Lefebvre counterposes social space against absolutist conceptions, against 'the homogeneous and isotropic space of classical (Euclidean/Cartesian) mathematics' (1991: 86). But, importantly, he does not reject absolute space, he merely asserts its ontological contextualisation in social space. This point is also valid for mental/representational space. Both mental and physical dimensionalities are present within the dialectic of social space. Soja explains:

> As socially produced space, spatiality can be distinguished from the physical space of material nature and the mental space of cognition and representation, each of which is used and incorporated into the social construction of spatiality but cannot be conceptualized as its equivalent.
>
> (1989: 120)

Thus for the historical materialist, mental and physical spaces represent important dynamics whose socialisation through material practice must be explained. That is to say, the objective[15] force of physical space, and the subjective, ideal dimension must not be considered in isolation either from each other or from the social space of material practice. Soja explains the conceptual challenge arising from this view:

> Defining these interconnections remains one of the most formidable challenges to contemporary social theory, especially since the historical debate has been monopolized by the physical–mental dualism almost to the exclusion of social space.
>
> (1989: 120)

Again, the dialectical quality of social space is emphasised in that it is seen to embody both mental and physical dimensionalities; both senses being fundamental qualities of material practice. For Soja, 'Spatiality is a substantiated and recognizable social product, part of a "second" nature which incorporates as it socializes and transforms both physical and psychological spaces' (1989: 129).

Having distinguished the materialist view of space from post-positivist alternatives, I want now to produce a summary view that brings the analysis back towards the question of embodiment.

The embodiment of space

Lefebvre argued strongly for the historical and cultural specificity of social space. Thus, only at the most abstract level can one refer to social space as a universal concept, for material reality testifies that 'We are confronted not by one social space but by many – indeed, by an unlimited multiplicity or uncountable set of social spaces which we refer to generically as "social space"' (Lefebvre, 1991: 86). Through the contiguity of time and cultures these social spaces are none the less interconnected.

> The intertwinement of social spaces is also a law. Considered in isolation, such spaces are mere abstractions. As concrete abstractions, however, they attain 'real' existence by virtue of networks and pathways, by virtue of bunches or clusters of relationships.
>
> (Lefebvre, 1991: 86)

This suggests that individual social spaces may be conceptualised and described by reference to ensembles of concrete abstractions, such as the core activities and sites through which humans produce social relations. Indeed, social space is nothing more than 'a specific space produced by forces (i.e., productive forces) deployed within a (social and determined/determining) social practice' (Lefebvre, 1991: 171). And such a space must embody,

> 'properties' (dualities/symmetries, etc.) which could not be imputed either to the human mind or to any transcendent spirit, but only to the actual 'occupation' of space, an occupation which would need to be understood genetically – that is, according to the sequence of productive operations involved.
>
> (Lefebvre, 1991: 171)

Therefore the unique character of each social space is informed by the manner in which each is *productively occupied*. This does not imply a narrow, economistic focus, but rather an inclusive consideration of the entirety of ways in which the production and reproduction of human needs is realised in a particular setting through local and endogenous material practices. These practices include the very broad range of human endeavours that produce social (and spatial) relations. Thus, the type of concrete abstractions which characterise a particular social space will depend upon the manner of its productive occupation in this very broadly conceived way.

For Lefebvre, the body is the 'productive occupant' of space. He maintains that 'it is by means of the body that space is perceived, lived, and produced' (Lefebvre, 1991: 162). In this he is at one with Eagleton who asserts that 'for Marxism ... it is that *eminently spatial object, the human body*, with which everything begins and ends (1988: xii) (emphasis added). Space is seen to be created by the social practices of lived, material bodies (this equates to Marx's idea of the sensuous activity of human social practice). Lefebvre demands that materialism recognise the body as the immediate 'site' for the production of space. As he sees it 'The whole of (social) space proceeds from the body' (Lefebvre, 1991: 405). Moreover, body and space, as with society and space, are portrayed in a mutually constitutive relation:

A body so conceived, *as produced and as the production of a space*, is immediately subject to the determinants of that space: symmetries, interactions
and reciprocal actions, axes and planes, centres and peripheries, and concrete (spatio-temporal) oppositions.

(Lefebvre, 1991: 195) (emphasis added)

Just as the body produces social space through material practice, so too does
encountered space play a role in the creation of social embodiment. As just one
possible example of this metabolism, feminist scholars have demonstrated the
power of socio-spatial organisation (in particular, the built environment of cities) to engender women oppressively. Recent work by Grosz has explored the
metabolism between urban space and the gendered body; this premised on her
belief in the 'historico-geographic specificity of bodies' (1992: 243).

The point of the immediately foregoing discussion has been to show that the
body and social space exist in a dialectical relationship – in other words, a relation
which is responsible for the experienced, phenomenal world of human beings.[16]

An embodied historical-geographical materialism

A conceptual outline

It now remains for me to draw together the various preceding analytical threads
and provide a historical-geographical account of embodiment. The account is
organised around two conceptual markers: first, Marx's idea of two natures; and
second, Lefebvre's concept of social space. I wish to stress here again that this
conceptual outline is framed at an abstract epistemological level, and in no way
represents a complete theory of embodiment. Indeed, there will never be a
single, undifferentiated theory of embodiment; rather, materialism seeks
contextualised understandings of how social embodiment occurred/occurs in
different times and places. The following account thus has a twofold purpose:
first to guide the empirical investigations that will render up these embedded
knowledges; and second, to suggest how certain spatio-temporal contexts may
be related in broader explanatory frameworks.

Two natures

A critical organising construct for a materialist approach to the body and space is
Marx's idea of first and second natures. Here, first nature is the organic field of
transformation which each society receives from its predecessor. This 'resource'
field must include the materials of both the physiological body and physiographical
space. This received field of nature is seen to be subsequently *socialised* through
human endeavour and thus take on its second historical form.

Hence, one may conceive of physiology and physiography as providing the
materials for the social spaces and the social beings of second nature. The
important consideration here is that the materials of first nature are organic

phenomena which provide the conditions for different socialisations. However, the resulting social forms of (second) nature are always culturally and historically specific. The historically and culturally defined socialisation of first nature is achieved through the agency of human labour. In the material activities of production and reproduction, human beings socially transform received physiologies and environments in historically circumscribed ways. As Harvey puts it, 'The production of space-time is inextricably connected with the production of the body' (1996: 276).

Importantly, first nature cannot be conceived as a set of ahistorical or immutable realities of form. First nature itself is open to transformation through both human intervention and internal evolution, though at a vastly slower rate than its socialisation. In addition, *human understandings* of organic nature are inevitably theory-informed. As a result, comprehensions of nature are bound to the movement of human thought in general, and must themselves be regarded as historically specific.

Merleau-Ponty (1962) insisted that bodies both produce, and are produced by, history. Lefebvre (1991) agrees and points to a similar relation between bodies and space. Thus it is important to think of the body and space as ever engaged in a historical metabolism:

> The conception of the human body (and all that goes with it – conceptions of self, subjectivity, identity, value, and social being) depends upon definitions of space and time. If the latter are relational rather than absolute, then it follows that conceptions of the body and the conceptions of spatio-temporality are mutually constitutive of each other.
>
> (Harvey, 1996: 248)

This mutually constitutive relation between embodiment and space–time foregrounds the role of the body itself in conditioning the production of human society. (Indeed, Dorn (1994) believes that a theoretical focus on embodiment can renew structuration methodology. He argues that the body constitutes a material intersection between structure and agency where the lived experience of power structures is revealed most clearly.) As Foucault laboured to show, the body emerges as a key site of political contestation in any society. The politics of the body – meaning the complex social contestation that attends its material signification – is a transformative force which can redirect vectors of social change, even at the structural level.

Of course, political contestation implies social resistance, evident in the myriad counterflows which oppose, and even repulse, the conditioning of everyday life by power structures. In this sense, there is always a politics of resistance centred on the body: 'The body ... is the first and most radicalized medium for resistance' (Dorn, 1994: 23). *Resistant bodies* simultaneously challenge and (through their very marginality) define hegemonic forms of power. This resistance lies at the heart of historical social change, especially, the adaptations, redirections and transformations that are forced upon power structures over time.

Finally, it is certainly possible to arrive at generalisations about how bodies are socio-spatially produced in specific epochs and places. The specific nature of this metamorphosis can, however, be identified only through empirical studies of particular socio-spatial contexts. Dorn comments:

> accepting the spirit of Marxism, emancipatory politics does require a conception of human potential which is trans-historical and trans-geographical, outside the play of representation ... A focus on embodiment does not necessarily abrogate the responsibility of geographers to inform grand, revolutionary theory. Nor should research on the body be seen as micro-regional geography. The body can receive blows from both the insensitive doctor and the commodified norms of personal attractiveness under Western capitalism ... this body ... can speak across space.
>
> (1994: 23)

Chouinard makes a similar argument, focusing on disabled embodiment:

> If critical geographies of disabling differences are to take into account the difference that different disabilities make, then they will need to be informed by critical, nonreductionist conceptualizations of the body and embodiment ... such theories will take seriously the notions that the body is inscribed ... in ways that empower and disempower, and that the material corporeality of the body makes a difference in how such processes unfold.
>
> (1997: 384)

Spaces of embodiment: two core considerations

Lefebvre has pointed out that social spaces are concrete abstractions which exist only in their cultural, and therefore conceptual, specificity. Accordingly, a critical question for the materialist contemplating the process of social embodiment must be: which space(s) is/are at issue?

Recognising that socio-spatial relations are embodied dynamics, and that social groups are diverse human collectivities, it is important to avoid the mistake of universalising the body across time and space. Empirical work must seek to understand how particular forms of embodiment were experienced in different times and places. Harvey demonstrates how this specification must inform analysis of real contexts, insisting, for example, that investigations of urban social space in capitalism must begin with the 'prior question ... of *whose* bodies produce the city versus *whose* bodies inhabit it' (1996: 278) (emphasis added).

As a socially and temporally bounded phenomenon, the concept of social space may be thought of as an attempt to describe the socialised metabolism between groups of bodies and territoriality. Here, biological bodies and physiography are held to be transformed through material practices into social beings and social spaces. The crucial point is that the practice of transformation can delimit the form of social being which certain physiologies may take. This delimitation is achieved through the creation of social spaces which constrain and/or devalue the identities of collectivities defined by specific forms of embodiment. An example of this is evident in Marx's portrayal of industrial capitalism as a social space that crushed and alienated the bodies of workers.

In summary, I suggest that materialist analyses of embodiment carefully specify both the social space(s) and the social group(s) within their empirical frames. These must be the methodological starting points for historical-geographical analysis. This book is concerned with one (oppressed) form of embodiment, disability. In the next and last section I will distil from the foregoing account of embodiment a framework for considering the socio-spatial production of disability. As I will demonstrate in the rest of this book, this framework can guide the study of disability in different historical-geographical contexts.

The socio-spatial production of disability

The natural basis of disability

How can the foregoing account of embodiment, and the specific analyses which informed it, be applied to the specific question of disability? To answer this question, I will now distil from the previous discussions a summary materialist framework for analysing disability. The obvious starting point for such a framework is the notion of 'two natures' which can be readily related to the materialist model of disability that was reviewed in the previous chapter.

In the materialist conception, 'impairment' and 'disability' correspond to first and second nature respectively. Impairment thus is simply a bodily state, characterised by absence or altered physiology, which defines the physicality of certain people. Importantly, no *a priori* assumption is made about the social meaning or significance of impairment. Impairment can only be understood concretely – that is to say, historically and culturally – through its socialisation as *disability* or some other (less repressive) social identity.

This is not to say that the materialist position ignores the real limits which nature, through impairment, places upon individuals. Butler and Bowlby (1997) are entirely right in their criticism of social models of disability which erase or immobilise the critical issue of embodiments (i.e., sex, gender, race, impairment) and their role in identity formation. Rather, the materialist view that I invoke here separates, both ontologically and politically, the oppressive social experience of disability from the unique functional limitations (*and capacities*) which impairment can pose for individuals. Impairment is a form of first nature that certainly imparts a given set of abilities and inabilities, which then places real and ineluctable conditions on the social capacities of certain individuals. However, the social capacities of impaired people can never be defined as a set of knowable and historically fixed 'functional limitations'. The capacities of impaired people are conditioned both culturally and historically and must therefore be defined in socially specific ways. Importantly, this social specification presupposes empirically informed analysis of how impairment was socialised and experienced in specific spatio-temporal contexts. Davis gives voice to this theoretical and methodological precept:

> In the task of ... theorizing disability, one of the first steps is to understand
> the relationship between a physical impairment and the political, social, even
> spatial environment that places impairment in a matrix of meanings and
> significations.
>
> (1995: 3)

Far from being a natural human experience, disability is what *may* become of impairment as each society produces itself socio-spatially: there is no *necessary* correspondence between impairment and disability. There are only historical-geographical correspondences which obtain when some societies, in the course of producing and reproducing themselves through cultural and political-economic practices, oppressively transform impaired first nature as disablement. The historical-geographical view recognises that different societies may produce environments that liberate the capacities of impaired people while not aggravating their limitations. In short, the historical-geographical approach opposes the *naturalisation of disability*, as an inevitable consequence of physiology, while insisting upon the *natural basis of disability*, as an oppressed form of embodiment.

It is certainly possible to point to historical societies where impairment was socio-spatially reproduced in far less disabling ways than has been the case in capitalism. The historical analyses of Davis (1995), Dorn (1994), Finkelstein (1980), Gleeson (1993), Morris (1969), Ryan and Thomas (1987) and Topliss (1979) have all opposed the idea that capitalist society is inherently less disabling than previous social forms. Davis enlists new historical evidence in his argument against naturalised accounts of disability:

> Recent work on the ancient Greeks, on preindustrial Europe, and on tribal
> peoples, for example, shows that disability was once regarded very differently from the way it is now ... the social process of disabling arrived with
> industrialization and with the set of practices and discourses that are linked
> to late eighteenth- and nineteenth-century notions of nationality, race, gender, criminality, sexual orientation, and so on.
>
> (1995: 24)

Several of the above analyses, including my own (Gleeson, 1993), have argued that while impairment was doubtless a prosaic feature of the feudal European world, disablement was not. This is a specific historical frame that I will explore in more detail in the next part of the book.

By showing that impaired people participated in the basic activities of previous societies, such studies falsify the 'Whig' history of disability.[17] Whig chronicles (e.g., Gordon, 1983) naturalise disability as a trans-historical 'tyranny of nature' which, thanks to the steady progress of technology and enlightened humanist practices – such as institutional 'care'! – is now all but conquered. By contrast, the historical-geographical position establishes impairment and disability in a temporally and spatially contingent relation that cannot be set within the fixed, historicist scheme of Whig history.

Geographies of disability

When it occurs, disability is both manifested and reproduced in socio-spatial ways. There will exist in such circumstances 'geographies of disability'. These geographies entail two overlapping domains of social relations. First, there are the socio-spatial patterns and relations through which impairment is oppressed by dominant power relations. Secondly, and just as importantly, there are the socio-spatial experiences and practices of impaired people who must negotiate disabling power structures in their everyday lives. Recalling the point made earlier about bodies of resistance, it must be recognised that these experiences inevitably include social actions and identities that counteract the oppressive flow of disabling relations. It cannot be doubted that these resistances are part of a broad social process through which disabling power structures are constantly renegotiated and changed. Dorn labels disability a 'dissident body', meaning a corporeality that is 'particularly resistant to articulated norms' (1994: 154). The sorts of norms he has in mind here include socially constructed ideals of beauty and physical aptitude.

While I fully support this foregrounding of dissidence as a key feature of disability, I think we must also be careful not to overstate its significance in the everyday experience of oppression. While many people who experience structural prejudice are able to practise small acts of daily resistance, larger forms of political dissidence may elude them precisely because they are marginalised and disempowered. It is important, therefore, to show how these 'quiet acts of resistance' congeal in certain times and places to become social movements that realise, to varying degrees, the potential of dissident bodies to make emancipatory changes. It would be an affront to reason (and evidence), for example, to suggest that the brutalised lives of disabled street people in mid-Victorian London (which I examine in Chapter 6) were significantly enriched by an active socio-political dissidence. But it is none the less true that these experiences of oppression in time helped to catalyse both a wider class anger at oppression and the eventual rise of disabled people's social movements (Campbell and Oliver, 1996).

Importantly, while these geographies describe social marginality – meaning *inter alia* physical exclusion and cultural devaluation – they do not imply socio-spatial experiences that are utterly divorced from the mainstreams of social life. Marginality and inclusion exist, through social power relations, in a mutually constitutive tension: thus there are no special 'worlds of disability', as some commentators (e.g., Golledge, 1993) have implied. Disability is a socio-spatial experience that emerges from core social relations. These sets of experiences – geographies of disability – must thus be mapped from the co-ordinates provided by the cultural, political-economic and spatial organisation of society.

Conclusion: histories that need to be written

My aim in this chapter was to outline a historical-geographical account of embodiment, and derive from this a framework for analysing disability in particular societies. To do this, my analysis sought to articulate the materialist debates on the production of nature and the production of space. Although historical-geographical commentators such as Smith (1984) and Harvey (1996) have also made this convergence previously, their purposes have differed from mine, which was to explore the socio-spatial production of embodiment. Thus my account placed embodiment at the heart of the material processes through which human beings transform received nature and thereby create unique social spaces. To recapitulate a crucial point, social embodiment is, like the material processes that produce it, sourced in specific historical-geographical contexts. Therefore, it is axiomatic that materialist analyses of embodiment carefully specify both the social space(s) and the social group(s) within their empirical frames. This specification, however, must not be an end in itself, but rather the first step in identifying historical-geographical continuities, echoes and divergences through a larger consideration of the linked evolution of human societies.

To repeat another vital point, the embodied materialist framework, and the derivative framework for disability, that I have presented are themselves just the first steps in a fuller theorisation of how impairment was/is experienced in different societies. As Dorn puts it (in considerably understated terms), 'The larger work of historical geographers on the place of disability in Western culture has yet to be pursued' (1994: 146). The key point of historical-geographical materialism is that all spatial analysts are, or at least should be, historical geographers, which is to say that a sense of the temporality of social space and embodiment is an essential, rather than an optional, feature of geographic enquiry. Indeed, as Harvey (1996) reminds us, critical scrutiny of the past is a political duty of contemporary historical-geographical enquiry.

When Foucault asserted that a 'whole history remains to be written of spaces … ' (1980a: 149), he clearly had in mind histories of embodiments in different cultural contexts, a task to which he contributed much through his studies of Classical Antiquity and modern European societies. Dorn also has forwarded an avowedly-materialist account of embodiment, framed within the historical evolution of the 'American capitalist space economy' (1994: 213). His study is impressive both for its historical sweep and for its analytical sophistication, highlighting a distinct, yet diverse and shifting, set of cultural and political-economic influences which attempted to fashion a heterogeneous national population into an army of 'rational' and 'productive' bodies. His study overlaps, and finds support in, Davis's equally ambitious investigation of how the emergence of capitalist modernity changed the socialisation of deafness in Europe. Both studies demonstrate the power of a historical-geographical approach to explain the emergence of disabling social relations in distinct cultural settings. Yet both studies also have empirical limits – neither, for example, draws upon primary materials. As Dorn explains, his aim was to produce through his theoretical analysis 'merely a skeleton allowing considerable fattening through empirical analysis' (1994: 222). He calls for new 'body histories' which will explain embodiment in 'particular regions' (ibid.), particularly seeking to expose instances where power structures were reproduced through the deployment of oppressive corporeal norms in cultural and economic life.

The rest of this book is devoted to this unfinished, indeed hardly established, project of writing 'body histories'. In the previous chapter I made it clear that there is a pressing need for empirically grounded research on the social experiences of disabled people in nearly all historical societies. In selecting the empirical contexts for this study from the vast continent of human history – including, for example, 'primitive' and Classical societies – I have followed a fairly well-trodden path. As was shown earlier, the transition from feudalism to capitalism has attracted interest from a variety of materialist and other radical disability scholars. Their reasons are clear enough: a distinguishing, and politically salient, feature of materialism is its insistence that the fundamental relations of capitalist society are implicated in the social oppression of disabled people. This suggests that the elimination of disablement (and, for that matter, many other forms of oppression) requires a *radical transformation*, rather than reform, of capitalism. Thus, *from the perspective of disabled people*, historical-geographical research is needed for two main reasons: first, to identify those specific and enduring features of our present social formation, capitalism, which oppress disabled people; and second, to demonstrate the ways in which impairment was experienced in alternative societies, with a view to identifying social arrangements that are non-disabling.

'Alternative societies' includes both societies which preceded capitalism and those which have existed alongside it. Given that capitalism has not been the exclusive source of disablement in human history, it is politically important that materialists turn a critical gaze towards the historical experience of disabled people in 'socialist' societies. I think it best, however, to begin the historical-geographical project from our present context, connecting back to its prior forms, to the societies which gave birth to capitalism. A better understanding of the historical genesis of disability in capitalism both serves the contemporary political needs of the disability movement and also establishes the basis later for meaningful comparisons with other real and potential modes of production. In the present volume therefore I intend to follow the interest of materialist disability scholars in contemporary and historical capitalism, and its antecedent social form, feudalism. My explorations begin with feudalism and early capitalism in Part II, and then shift in Part III of the book to contemporary themes.

Part II
Historical geographies of disability

4 Historical-geographical materialism and disability

Introduction

The argument made up to now has been that the social position of disabled people in any society can be fully understood only through socio-spatial analysis of lived experience, past and present. My main interest in this book is in the Western experience of disability, which demands consideration of how capitalism has shaped the lives of physically impaired people. In order for this clarification to occur, it is necessary to contrast the experiences of disabled people in capitalist and non-capitalist societies. There are two broad ways of framing this comparison – i.e., historically and cross-culturally. In this book I wish to contrast Western capitalism with its antecedent social form, feudalism.

My reasons for choosing this set of empirical frames are both personal – I have long been interested in everyday life within pre-modern Europe – and analytical. The analytical ground is that studies of historical change within defined social spaces, including an entire mode of production, can achieve an empirical and conceptual coherence which sometimes eludes cross-cultural studies. This coherence in turn provides a powerful basis for the explanation of historical phenomena, such as the shifting fortunes of a distinct social group.

Any historical analysis must have both an empirical starting point and a framework for understanding temporal social change. Disabled people's lives have been shaped and differentiated by the historical structuring of social relations around a variety of social cleavages, such as class, gender, race, and sexuality. I chose a political-economic frame for my empirical analyses because the historical rise of capitalism generated a profound, and inescapable, source of material change in the increasing array of societies which have yielded in time to commodity relations. I therefore wish to elaborate how this vital transformative force has affected the social geographic circumstances of disabled people. By choosing a political-economic historical framework, I do not wish to dismiss or downplay the contribution of other socio-cultural structural influences on the historical experience of disability. My analyses will capture at various points the effects of these other socialising forces. None the less, I cannot, and do not, claim that the following historical geographies provide a complete picture or explanation of the changing experiences of disabled people in Western societies. It will be the task of subsequent historical geographies of disability to elucidate more fully the complex influences of various identity forms on the past lives of disabled people. I hope to contribute to this process by offering glimpses on the role that political-economic dynamics played in this historical process.

The aim of this chapter is to provide a conceptual introduction to the second part of the book which deals with the historical experience of disability in feudalism and industrial capitalism. In this chapter, I will distil from the framework developed in the previous part of the book a set of historiographical principles which can guide the study of disability in past societies. To do so, I will first need to engage the present historiography of disability which I referred to briefly in Chapter 2. My analysis here will begin with a short, critical review of this historiography, followed by an outline of my alternative historical-geographical method of analysis.

Conventional approaches to the history of disability

As I explained in the previous two chapters, there has been relatively little attempt within social science to understand the historical experience of disability in any depth. The few serious historical studies of disability hardly constitute a comprehensive and critically engaged debate on the topic. Moreover, the limited historiography of disability studies seems to have littered the field with a number of assumed orthodoxies about the social context of impairment in previous societies (Gleeson, 1996b). I want to examine these assumptions critically in the following discussion with a view to providing an alternative historiography in the second part of this chapter.

Orthodoxy one: disabling ideologies

The first orthodoxy is the belief that powerful 'disabling ideologies' – including religious and philosophical outlooks – in pre-modern European societies were directly responsible for the historical oppression of impaired people. Smith and Smith, for example, point to,

> the Judeo-Christian ethic of associating physical defects with sin. Since people are supposedly created in the image of God, anything which fails to fit that image is deemed imperfect – that is, not Godly – and hence evil. According to this judgement, people with physical disabilities, through their obvious blemishes, are wanting and epitomised as bad.
>
> (1991: 41)

This historiographical orthodoxy is commonly associated with disability chronicles that rely mostly for empirical evidence upon sacerdotal texts and prescriptions. Winzer's (1997) essay on disability in the pre-modern era is an example of this approach. In spite of an opening declaration on the need for histories rooted in analysis 'of daily life in these cultures' (1997: 75), Winzer's account draws principally upon surviving literary and religious texts, and barely attempts to reconstruct the quotidian contexts within which both ideology and law were practised and disability was experienced.

Two objections may immediately be raised to this orthodoxy and its historiographical implications. First, it is not at all clear that disabled people were subject to *universal* social or religious antipathy in pre-modern societies. This is an *a priori* speculation which ignores the complexity of how discursive religious and ethical mores were socially enacted for populations, including 'special social groups' such as disabled persons. The tendency to read historical material reality directly from ideological/religious texts or aesthetical records of the past is a failing of idealist approaches in general. Dorn, for example, argues that in the feudal era, 'there is little evidence that impaired or abnormal bodies were set aside as a separate category' (1994: 20).

Second, this conjecture is sourced in a very limited methodological outlook that supports a theoretical simplism – the 'Judaeo-Christian ethic' – to justify a failure to consider the possibility of complicating historical realities. The history of Judaeo-Christian thought *and practice* can hardly be explained through appeal to a single 'ethic'. Christianity had a much more complex presence in European society than such a construction would allow, with its evolving teachings subject to localised interpretations in varying periods, ranging from fervent (if often sporadic) devotion to outright rejection. It is reasonable to assume that much of the quotidian social experience of Christianity in the feudal era was marked by loose, if genuine, observation that embraced both moments of strict adherence and permissiveness. Moreover, as Harvey (1996) explains, pre-modern 'Christendom' was composed of social spaces heavily coded by religious and moral concepts that related in diverse ways to the daily routines of individuals: 'The result was a variety of spatio-temporal conceptions deriving from different modes of experience (agricultural, political, ecclesiastical, military, etc.)' (Harvey, 1996: 214).

Even theologically, Judaeo-Christian thought was hardly a cohesive 'ethic', being characterised by discrepancies of interpretation at many levels; the constant disagreements over the spiritual significance of material phenomena, such as bodily differences, being one example of these. There were certainly many lines of religious thought on the question of disability. The influential philosophy of Spinoza (1632–77), for example, opposed negative constructions of disability. For Spinoza, 'A physical ... cripple is such because of its place in the system: God has not tried to produce perfection and failed' (Urmson and Ree, 1989: 305). In addition, in the realm of everyday life, feudal peoples may have welcomed the presence of disabled mendicants, as Braudel explains: 'In the old days, the beggar who knocked at the rich man's door was regarded as a messenger from God, and might even be Christ in disguise' (1981: 508).

Though subject to a variety of interpretations (e.g., Bovi, 1971; Foote, 1971), the inclusion of various groups of lame beggars in the works of Bruegel (1520?–69) – especially *The Battle between Carnival and Lent* and *The Cripples* – would seem to signify that those with physical 'maladies' had a place *within* the pre-modern social order.

Figure 4.1 The Battle between Carnival and Lent (Pieter Bruegel); Kunsthistorisches Museum, Vienna

Consider for a moment *The Battle between Carnival and Lent* (Fig. 4.1) which depicts many of the human elements of the medieval social order. The painting gives us a kaleidoscopic view of the European feudal landscape with its rich portrayal of social characters engaged in horseplay and work within a town square. Within this panorama of gambolling figures are the lame beggars shown in the detail (Figure 4.2). This group is set rather unremarkably within the larger, symbolic *mise-en-scène*, and Bruegel seems to be telling us that physically disabled people were very much a part of the feudal social order.

Of course this sort of reading of what is merely one 'historical text' can only be suggestive, but it does help to problematise the negative disability histories that rely solely on surviving religous or philosopical records.

Orthodoxy two: the beggared view of history

The other orthodoxy evident in many histories of disability is the view that all impaired people were beggars in the pre-industrial era. This assumption is explained by Safilios-Rothschild:

> the disabled *have always* been 'problematical' for *all* societies throughout history, since they could not usually perform their social responsibilities satisfactorily and became dependent upon the productive ablebodied (1970: 12).

> (emphasis added)

Figure 4.2 Detail of *The Battle between Carnival and Lent* (Bruegel)

Hahn is also convinced that disabled people in the pre-modern world were doomed to become either beggars or minstrels 'who wandered through the countryside until they became the first group to receive outdoor relief under the English Poor Law of 1601 and subsequent legislation' (1988: 29). Elsewhere he repeats this view in even more strongly fatalistic terms:

> To the extent that disabled persons had any legitimized role in an inhospitable environment prior to the advent of industrialization, they were beggars rather than competitive members of the labor force.
>
> (Hahn, 1987a: 5)

Consequently, 'Unlike most disadvantaged groups, disabled adults never have been a significant threat to the jobs of nondisabled workers' (Hahn, 1987a: 5).
More recent evidence of this orthodoxy is supplied by Winzer:

> In the thousands of years of human existence before 1800, life for most exceptional people seems to have been a series of unmitigated hardships. The great majority of disabled persons had no occupation, no source of income, limited social interaction, and little religious comfort ... Their lives

were severely limited by widely held beliefs and superstitions that justified the pervasive prejudice and callous treatment. Individuals seen as different were destroyed, exorcised, ignored, exiled, exploited – or set apart because some were even considered divine.

(1997: 76)

The effect of this historicist orthodoxy is to silence history, projecting disabled people's relatively recent experience of service dependency and marginalisation through the entirety of past social formations. This assumption must be rejected on two grounds. First, it is based on a limited reading of extant textual and visual records of disability and makes no attempt to capture the concrete experiences of impaired persons in historical societies (Scheer and Groce, 1988). The view of all disabled persons as beggars is based upon a very limited empirical appreciation of life in past societies, thus raising the inevitable question of reification. (Philo (1997) presents a compelling argument against the stereotypical image of the 'medieval mad person' as an excluded Other, supposedly subjected to blanket exclusion from mainstream communities.) Second, this construction of disability in history has odious political implications by naturalising the relationship between impairment and social dependency that has existed to varying degrees in capitalist societies.

The historicist tendency to 'beggar' the entire history of disability is revealed in Stone's (1984) major chronicle of Western public policy towards impaired people. I think it worthwhile to explore Stone's account in some detail here, both because of its influence in disability studies, and also for the manner in which it clearly exhibits the shortcomings of the historicist orthodoxy.

As its title – *The Disabled State* – indicates, Stone's (1984) history is predicated upon a statist approach.[1] In this she posits the historical existence of dual 'distributive systems' in societies: one involving the activities of those producing sufficient value to meet their own needs and more; and the other, what may be described as a 'social circuit of dependency' which includes those who cannot maintain self-sufficiency. From this dualism a basic 'redistributive dilemma' is held to arise, presenting an enduring socio-political problem for states. The tension between the two systems based on work and need is the *fundamental distributive dilemma* (Stone, 1984: 17) (emphasis added). For her, disability is explained as a juridical and administrative construct of state policy which is aimed at resolving this supposed redistributive predicament.

I have two general criticisms of Stone's chronicle. First, the historiography of the account is both selective and ambiguous. The chief defect is the projection of the 'redistributive dilemma' construct seemingly through all history; an epistemological presumption which has little empirical substance. This 'distributive dilemma' is, for example, of doubtful relevance to the explanation of primitive societies where a dichotomy between 'producers' and 'dependants' was neither obvious nor culturally enshrined.

In reality, Stone is referring to a far more recent episode of human history where social formations have been characterised by remuneration systems which assume a direct reciprocity between *individual* work and *individual* reward. That

Stone really has these social formations in mind is evidenced by her claim that 'societies' 'face the problem of how to help people in need without undermining *the basic principle of distribution according to work*' (1984: 15) (emphasis added). The reciprocity between work and reward for individuals which is assumed here is not a 'basic principle' in primitive societies. Mandel (1968: 31) provides clarification on the primitive organisation of labour: 'Differences in individual productive skill are not reflected in distribution. Skill as such does not confer a right to the product of individual work, and the same applies to diligent work.' The co-operative character of the primitive labour process favours a communal, rather than individual, distribution of the social product.[2]

The anthropologists Dettwyler (1991) and Scheer and Groce (1988) find little empirical support for the idea of a 'distributive dilemma' in any past society, let alone in primitive social forms. Dettwyler (1991) sees the social category of dependency as exceedingly fluid, and warns against the tendency to reduce it to physical impairment: 'In reality, every population has members who are, for varying lengths of time, nonproductive and nonself-supporting' (1991: 379). This author believes that 'as with children, disabled people in most societies participate as much as they can in those activities that they are capable of performing' (1991: 380). Thus, 'Every society, regardless of its subsistence base, has necessary jobs that can be done by people with disabilities' (ibid.). The consequence of this view is that 'It is presumptuous of anthropologists to assume that they can accurately assess how productive disabled individuals might have been in the past' (1991: 381). One would expect the accuracy of such analysis to be rather better for societies in the more recent past; Dettwyler is probably thinking of primitive society when making this remark. But the comment serves as a general caution against the historicist tendency to naturalise the idea of a 'distributive dilemma' in human history.

By assuming that certain modes of production shared universal qualities, Stone (1984) is led to adopt confusing generalisations, such as seemingly equating 'peasant' societies (a vague term in her analysis) with subsistence forms of production. A subsistence community is characterised by the absence (or extreme limitation) of productive surplus and most commonly refers to simple societies such as tribes or hunter-gatherer groups (Jary and Jary, 1991). Peasant societies, by contrast, embody a different form of social development, usually organised around an agrarian economy, and where surpluses may be both common and significant. Consequently, Stone's (1984) analysis must be seen as applying only to relatively recent Western modes of production – namely, feudalism and capitalism – in spite of the wider historical ambit it assumes.

The second objection to Stone's (1984) account is that it avoids or trivialises the central motive force of distribution – the social relations of production. The statist approach emphasises disability as a juridical and administrative construct, thereby dematerialising the social context of impairment. This approach can only reveal the meaning of disability *to the state*; it cannot adequately claim to capture the concrete reality of impairment within social relations generally. The juridical record, in particular, cannot divulge to us the historical lived experience of disabled people, however much the law may have helped to shape the social context of impairment.[3] The primary motive force in the social construction of disability is the cultural material organisation of production and reproduction. Disability, as a policy response of states to the contradictions of exploitative modes of production, is itself a cultural material force in social relations. Thus, the actual lived experience of impairment in the past can only be sensed through cultural materialist analyses of the organisation of production and reproduction.[4]

The great danger of chronicles such as Stone's (1984) is that they encourage belief in a 'beggared' history of disability. The tendency is to reduce the concrete lived experience of impairment to the more limited domain of disability as state social policy. This reduction can only cloud the material genesis of disability and reify the state dependency that has overshadowed the social experiences of many disabled people in recent and contemporary capitalist societies. The histories of disabled people, marked both by shifting materialities and by socio-spatial differences, are reduced thus to a single saga of vagabondage and marginality.

A final, though critically important, objection to the orthodoxies explored above is that they both tend to deny any agency on the part of disabled people in past societies. The 'beggared' history view in particular reduces disability experience to that ordained by ideological structures, thus failing to appreciate the ways in which individuals – indeed, even entire social groups – negotiate, subvert and resist power structures in the material realms of everyday life. As Philo (1997) reminds us, totalising stereotypes of difference – historical or contemporary – neglect the 'messy realities' that emerge when oppressive structures are applied to socially and geographically diverse societies.

A historical-geographical approach

Having critically reviewed aspects of conventional historical accounts of disability, I want to now outline briefly an alternative historiographical approach, drawn from the historical materialist framework that was explained in the previous chapter. As will become evident, I am, like many other critical geographers, attracted to the historical methods deployed by theorists of the *Annales* school, both for its close resonance with historical materialist thinking and for its attentiveness to spatiality.

The historical comparison of social spaces

The previous chapter outlined an embodied materialism, and from this, a socio-spatial framework for analysing disability. This theoretical outline ended with a discussion of 'histories that need to be written', emphasising the need for empirically grounded research on the socio-spatial experiences of disabled people in nearly all historical societies. To give this very broad empirical remit a methodological focus, I pose the following research question for new histories of disability:

> *How have changes in the socio-spatial organisation of societies affected the lived experience of physical impairment?*

This specific query reflects the basic premise laid down in the historical-geographical account of embodiment; namely, that the experience of embodiment is shaped by the specific, though always changing, socio-spatial organisation of society.

This research question presents a general methodological imperative: the need to examine the experience of impairment within clearly discrete socio-spatial settings. It will immediately be seen that this methodological requirement invites both diachronic (historical) and synchronic (cultural or cross-cultural) comparisons of specific social contexts. However, as I stated in the introduction to this chapter, my main social comparisons of disability experiences in this book will be historical. The third part of the book will involve some contemporary comparative analysis, though this will be rather closely framed within the confines of contemporary Western societies.

Confining the comparative social analysis of disability in this book to *contemporary Western* countries would reduce my ability to explore the significance of core capitalist relations for disabled people. My historical analysis will therefore span the transition between two historical landscapes, or 'modes of production' – feudalism and capitalism. As I pointed out in Chapter 2, this period of transition has been of greatest interest to materialist disability theorists undertaking historical comparisons of the socialisation of impairment.

This is not to deny that contemporary capitalist societies exhibit significant socio-spatial differences which give rise to distinct social contexts for disability. These distinctions between disability experiences in various capitalist countries are certainly worthy of geographers' attention – a point I will return to in the concluding chapter.

Historical method

Having indicated some broad methodological and empirical settings for my analyses, what methods seem best suited to historical-geographical enquiry into disability? The historiographical approach I wish to take here is broadly informed by that of the *Annales* school, particularly that evidenced in the writings of Bloch (1962, 1967) and Braudel (1973, 1981).[5] The chief innovation of the *Annales* historians was to leaven the diachronic construction of historical accounts with the synchronic understandings of other social sciences, such as Geography, Political Economy and Sociology (Jary and Jary, 1991). Their concern was with the tendency of historians to produce histories,

> in which dramatic events and picturesque individuals follow one another all too loudly across the stage, to the exclusion of those years and classes of society too uneventful or humble to meet its criteria of what is actually 'historical'.
>
> (Sturrock, 1986: 60)

Against such 'event-histories', the *Annales* scholars counterposed realist chronicles of the everyday (the 'uneventful') world, underscoring the fact that societies are complex socio-spatial phenomena. Importantly, the everyday – or 'quotidian' – aspects of the past are held to have emerged from people's lived experience of social structures, such as the feudal peasant economy. Accordingly, the *Annales* historiography achieves its object of synchronically informed accounts of the past,

> by attempting to reconstruct the permanent or more or less permanent constraints within which all these events occurred, and also ... by concerning itself with the whole of contemporary societies instead of just with those elements of it fortunate enough to have singled themselves out.
>
> (Sturrock, 1986: 60)

The historiographical significance accorded social structures by the school is nowhere more apparent than in Braudel's encyclopedic study of the quotidian human experience of early capitalist Europe (1400–1800).

This dual historiographical emphasis on daily life and the cultural material structures which shape it can be seen in stark contrast to the orthodoxies identified in the previous section. The *Annales* approach rejects the idea that a social group's historical fortunes can be understood through simple resort to surviving records of the ideologies and laws of ruling interests. Importantly, this approach is far more likely than those reviewed earlier to reveal the complexity of human responses to structural forces, including political economy, religious ideologies and edicts, and laws. As social scientists (e.g., Miller *et al.*, 1997), including geographers (e.g., Pile and Keith, 1997), are well aware, structures condition rather than determine human social life, and individuals negotiate these larger forces in everyday life through a variety of socio-spatial strategies. These quotidian strategies can be located on a broad spectrum of possibilities ranging from active resistance to enthusiastic acceptance, with perhaps the bulk of human experience falling somewhere between, characterised by various shades of everyday endurance (Laws, 1994). There is every reason to believe that disabled people's responses to oppressive structures in the past included many strategies within this range – indeed, the two historical studies which follow this chapter will emphasise this point.

Having established both the need for a historical empirical study of disability, and the methodological basis for this, the question of periodisation arises. What are the most appropriate historical contours for the study? As stated earlier, my political economic interest in disability informs my preference for 'mode of production' as a concept that can be used to structure analysis of the past.

Mode of production and social space

Historical-geographical theory stresses mode of production as a construct which lays the basis for a division of human history into different stages of social development. More specifically, Marx used the concept to distinguish between epochs of history in which different modes of producing (and reproducing) material life prevail. At issue are the forces of production (the human capacity to transform nature)[6] and the relations of production (ownership and control of the productive forces). The specific form of each, and the type of metabolism between them, were held by Marx to have defined a mode of material life (Godelier, 1978; Harvey, 1982; Bottomore *et al.*, 1983). In this way a distinction is drawn between capitalist, feudal, 'Asiatic', etc., modes of production, corresponding to historical social forms which structured everyday life in distinctive ways.

McQuarie (1978) is right to remind scholars using the mode of production construct that it is an abstraction which refers to a set of relations predominating within, rather than exclusively defining, a given social formation. Movement between the historical dominance of various modes is, as Marx insisted, neither sudden nor always transparent.[7] Thus, while within each epoch so defined a given set of social relations will overshadow all others, various modes may none the less be present, lingering from the past, or germinating as new forms. McQuarie provides concrete historical examples: 'Next to slave-worked latifundia of the Roman aristocracy existed small peasant farms; next to the medieval manors, factories and artisan workshops thrived' (1978: 28). This diversity of modes of production within social formations has a spatial logic. The preponderant set of relations will, of course, occupy the principal social spaces in which mainstream production and reproduction activities are undertaken, while other modes will tend to breed and wither in relatively marginal or liminal terrains.

Mode of production thus provides a spatio-temporal framework for historical analysis. Lefebvre (1991: 31) has commented that 'every mode of production ... produces a space, its own space'. For him, 'the shift from one mode to another must entail the production of a new space' (1991: 46), though it must be stressed again that this new social terrain can never be completely universal, but must be co-extensive with smaller new and remnant forms. The move to a new mode is sourced in antagonisms within the social relations of production and reproduction (Marx, 1978), a transformation which Lefebvre (1991) sees as initiating new spatial practices. The new mode of producing and reproducing material life is nothing less than a set of practices which produce a new social space.

As Harvey (1996) observes, such a 'modal' shift to a new social space provokes change in existing structures of difference based, for example, on class, gender, race, religion and sexuality. This complex process is hardly uniform in time and space and is best viewed as a set of 'continuous and often contradictory movements within the historical geography' of social development (1996: 320). New modal spaces, as they arise, produce new patterns of domination and resistance – Harvey gives the example of how early capitalist political economic relations transformed gender and racial oppressions in Europe and its colonies, producing a sophisticated and diverse system of socio-cultural controls which aimed to dominate nature and certain forms of human embodiment, notably female and non-white corporealities:

> For example, the bourgeois tactic of depicting some segment of humanity
> (women or 'the natives') as part of nature, the repository of affectivity and

therefore disposed to be chaotic, 'irrational', and unruly, allowed those seg-
ments to be subsumed as elements requiring domination within the general
capitalistic project.

(1996: 320)

Thus, it can be seen that a modal shift signals a potentially profound change in
the course of social embodiment, involving new forms of freedom, prestige and
wealth for some, and new types of restraint, discrimination and deprivation for
others.

The social space of disability

How can the socio-spatial experiences of a particular social group, such as disabled
people, be elaborated for comparative purposes within various modes of production? Lefebvre (1991) has suggested an abstract typology for a materialist conception of social space that can be used for this purpose. His construct embodies
a triad of conceptual 'sites' – the 'public', 'intermediate' and 'private' nodes of
social practice – around which an analyst might compose a portrait of everyday
life for a social group.

Lefebvre seemed content to leave this construct, and each constituent element, as a loosely defined abstraction, a starting point for concrete analyses of
particular social spaces and their users. The implication is that the analyst will
specify the model in use. Accordingly, the following comments are intended to
fine-tune the model so that it can be used to frame empirical analyses of the
historical-geographical experiences of disabled people. I refine the model by
specifying the three elements ('public', 'intermediate' and 'private') in turn, in
order to reflect key sites for disabled people in both the feudal and early capitalist landscapes.

For Lefebvre, the 'public' node refers to a heterogeneous group of sites, including temples, palaces, administrative areas and political centres. For my purposes, the public focus will be on institutions; in other words, any place of communal
confinement and/or restrictive 'care'. My definition of 'the public' is admittedly
a restricted one, certainly narrower than Lefebvre's. The lack of surviving empirical records (see Appendix) makes it difficult, if not impossible, to explore the
presence of disabled people in other public spaces, such as the village square
depicted by Breugel. There are, however, records of institutions, public spaces
that would have been of critical significance for the lower orders in both study
periods. In the feudal era, monasteries, leprosaria, gaols and the village poorhouse represented important institutional forms for the peasantry, while the equivalents during industrial capitalism were workhouses, hospitals, prisons, asylums,
and other more specialised facilities.

The second class of sites in the model are identified as 'intermediate', meaning,
mostly, spaces set aside for circulation and commerce. The category here is re-
fined to mean workplace, or any site of production, thus stressing a pivotal material location for lower social strata.

The third category in Lefebvre's typology is named as 'private', by which he means the residential domain, broadly defined. Little immediate specification is needed here, though I will refer to this site as 'home', as the notion of privacy invoked by Lefebvre is a modern one with only limited applicability to medieval domestic space.

Thus, my specification of Lefebvre's general typology as a 'social space of impairment' embodies three key nodes: institution, workplace and home (Figure 4.3).

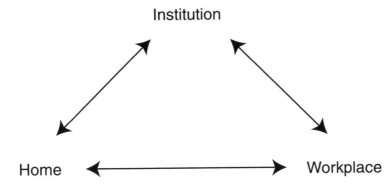

Institution

Home   Workplace

Figure 4.3 Typology for a 'social space of impairment' in feudal and early capitalist eras (amended from Lefebvre (1991))

As with all abstractions, the construct embodies a very general picture of concrete social history, and loses considerable detail in the process. The main difficulty lies in the extent of history from which the typology attempts to abstract.

The conceptual separation of home and workplace for both the feudal and capitalist historical periods is especially problematic. Only the concrete socio-spatial conditions of industrial capitalism provide unqualified support for this abstract dichotomy. The social space of the feudal peasantry was characterised by a relatively intimate union of domesticity and labour (Mumford, 1961). This intimacy of domesticity and labour is further distinguished from the industrial capitalist case by the fact that work for pay, or some other reward, was a minor part of the peasant household's productive activity. The peasant household was, to some extent, a self-sustaining economy with limited exogenous connections.[8] Work and domestic life combined without the formal distinction between paid ('productive') and unpaid ('reproductive') work spheres that characterised industrial capitalist social relations. Though slowly dissolving from around the sixteenth century onwards, this spatial harmony was decisively shattered by the Industrial Revolution which eventually established an almost universal separation of domestic and paid productive spheres (Buck, 1981; Mackenzie and Rose, 1983)

This uneven applicability of part of the model will not, however, thwart its general usefulness here as an analytical construct of social space in both historical settings. The common assumption that home and work activities were co-extensive in feudal social space is accurate in a *relative* sense (i.e., *vis-à-vis* industrial society), but its uncritical adoption blinds historical social analysis to subtleties in the make-up of the medieval landscape. The productive activity of peasant house-holds ranged (to varying extents) well beyond the family cottage. Men's work tended to be carried out in areas external to the home (but not far from it – in other words, manorial lands, family tenements), while women's productive activities centred around the cottage (Hanawalt, 1986). Thus, the home–work dichotomy may be usefully deployed in a critical analysis of the medieval setting as it keeps in view both the gendered division of feudal labour, and the fact that peasant crofts were the centre, rather than the limit, of productive activity.

Social focus

My historical analyses in this book will centre on the lived experience of impairment within the subordinate social classes of feudalism and capitalism. This decision is made for two reasons. First, it reflects my politically informed interest in the social experiences of oppressed peoples and classes. A first, and admittedly modest, step in liberating oppressed peoples is taken when their historical and contemporary experiences are themselves liberated from the crushing silence of official neglect or censorship.

Second, the historical experience of physically impaired persons within the subordinate social classes has not been explored in any depth, and this presents an important new field of research for historical-geographical materialism. The tendency of the oppressed to feel the greatest effects of socio-spatial change makes this experience an important one for the progressive social scientist.

As will be seen, I am also keenly interested in the effects of gender structures on the historical-geographical experiences of disabled people. Feminist historical-geographic analysis (e.g., Mackenzie and Rose, 1983) has explained how the transition to capitalism was achieved through a combined reshaping of class and gender structures. The consequences of the new capitalist gender structure – Patriarchy – for disabled people were surely profound, though there has been little attempt to explore this issue in disability studies. I hope my analyses shed some light on the historical significance of Patriarchy for disabled women and men.

Conclusion

In this chapter I set out to appraise critically certain historiographical orthodoxies that have pervaded the limited, though now growing, historical literature on disability. The historical-geographical approach outlined in the previous chapter undermines several of the key assumptions that have characterised much of this literature. I focused in this analysis on two problematical orthodoxies, the 'disabling ideologies' thesis and the 'beggared' history view. Both orthodoxies tend to produce a historicist view of impairment, as a universally oppressed form of embodiment. Such an assumption contradicts the dynamic social ontology of the historical-geographical view, and is also at odds with contemporary empirical and theoretical understandings of how structures are experienced in everyday life.

In contrast to this historicism, I proposed a historical-geographical methodology for the study of disability, drawing freely upon the approach of the *Annales* scholars. This methodology combines an appreciation for the abstract structures that condition social action with an empirical emphasis on how these forces are experienced by individuals in various ways, and in different places. My approach also stressed the concept of mode of production as a device for structuring historical analysis based upon key shifts in the political economy of nations and regions. In concrete terms, mode of production describes a set of distinct social spaces that are none the less conditioned by a set of common political economic arrangements. My historical interest in this book is in feudal and industrial capitalist societies, with a view to exposing how political economic forces, in conjunction with other structures, shaped the everyday lives of disabled people among the 'lower' social orders.

Having elaborated some basic historiographical settings for research on past disability experiences, I now turn to an empirical examination of two case settings, feudal Europe and industrial capitalism. As will be explained, a range of methodological considerations leads me to focus the empirical content of these two case studies on the English experience of feudalism and industrial capitalism. In the latter case, I will also draw empirical materials from another industrial capitalist context, colonial Melbourne.

5   The social space of disability in feudal England

Introduction

This chapter explores how the impaired body was socialised – that is, lived *socio-spatially* – in feudalism. The empirical analysis is rooted in the English medieval context. Broadly, the term 'medieval England' refers to the period between the Norman Conquest (1066) and the sixteenth century, during which time a feudal mode of production may be said to have prevailed (Bloch, 1962; Anderson, 1974a).

The empirical focus on the English feudal experience establishes a useful conceptual and spatio-temporal connection with the second case study, industrial capitalism (Chapter 6). England quickly freed itself from the socio-political bonds of feudalism, and provided centre stage for the Industrial Revolution. It is thus a commonplace to regard England as the cradle of industrial capitalism. Indeed, recognition of this fact was a major reason why Marx concentrated his economic analyses of capitalism on nineteenth-century England, where the rapid transition to factory production had propelled the early development of bourgeois social relations.

This study will not cover the entirety of the feudal landscape. As I made clear in the preceding chapter, my historical interest here lies with the experience of impairment within the 'lower orders'. Feudal England was only weakly urbanised, and the vast majority of its peasantry – probably in the region of 90 per cent – lived in the countryside, mostly in agricultural villages and hamlets (Anderson, 1974a; Hilton, 1985). Thus, I set the empirical frame for this investigation around the everyday experience of disabled people in rural areas.

There are few surviving historical records of the peasant domestic sphere in feudalism. This historiographical fact makes peasant life a difficult terrain for historical-geographical analysis. The silence of historical sources has, however, not prevented feminist and other historians such as Brooke (1978), Lucas (1983), Hanawalt (1986) and Labarge (1986) from attempting to reconstruct the everyday experience of peasant women in feudal societies. In the face of scant and disjointed empirical evidence, the approach taken has been to infer the gendered characteristics of everyday labour from the known features of the peasant household.

As with gender, impairment seems to have escaped the eye of the medieval chronicler, only more so. Most physically impaired people in feudal England were doubtless submerged within the peasant masses who, as the *Annales* scholars would have it, lived lives 'too uneventful' to warrant mention in the chronicles generated by clerics, nobles and state officials. But, as gender-conscious medievalists have shown, this difficulty need not preclude analysis of the position of particular social groups within the empirically amorphous peasant masses. In keeping with the *Annales* approach, my investigation of impairment proceeds by first revisiting the major settings of feudal peasant life. Impairment is explored through a deductive evaluation which takes the known socio-spatial structures of medieval life as its premises and, from this, infers the possible limits of the experience of impairment. The movement of analysis is from the general settings of feudal life (the manor and the village) to the immediate 'life spaces' of peasants (home, workplace and institution).

In the previous chapter I presented an amended version of Lefebvre's three-way typology for social space, an investigative tool that can be used to elaborate the everyday historical experience of disabled people. In this and the following chapters, my choice of empirical materials will be guided by this conceptual typology. Thus, in the following case study of feudal England, I have drawn upon a range of primary and secondary sources which reveal something of the character of home, workplace and institution in everyday peasant life. My data sources include the poor law surveys of Norwich (1570) and Salisbury (1635), both of which reveal glimpses of the social roles assumed by some disabled people in the middle ages. In order to save the reader from potentially distracting technical discussions in this and the following study, I have assembled in the Appendix my detailed explanations of the historical sources used in the book.

The chapter has two main parts. The first sketches the broad material context of peasant society in feudal England, emphasising the manor and the village as the principal terrains of everyday life. After this, I investigate the social space of impairment within these peasant settings by focusing on the three sites, home, workplace and institution.

The material context of everyday life

The political-economic context

By the end of the first millennium, the feudal state had become the chief form of political and social organisation in Western Europe. Anderson identifies two central features of feudal government: 'It was a state founded on the social supremacy of the aristocracy and confined by the imperatives of landed property' (1974b: 41). An important distinction between feudal and capitalist modes of production is the absence of any formal separation between political and economic power in the former. Anderson (1974a) describes the feudal order as a juridical compound of economic exploitation with political authority. The political order of feudalism was integrated by a chain of dependent tenures through which the economic surplus was initially extracted from the peasant producers, and subsequently divided among a hierarchically structured ruling class, made up of lower and higher nobles, who were bound to provide fealty, and sometimes financial support, to a monarch.

The central feature of the feudal mode of production was its dominance by the land and a 'natural economy', in which neither labour nor its products were commodities. Important here is the fact that capital – as self-expanding wealth – was hardly present in feudal society (Hilton, 1975; Le Goff, 1988; Wallerstein, 1983). The producing class of feudal society – the peasantry – were tied to the means of production – principally, the land – by a specific social relationship, serfdom (Anderson, 1974a). Serfdom – legally denoted as *glebae adscripti* (bound to the earth) – was a lawful circumscription of peasant social space, usually taking the form of a prescribed restriction on an individual's place of residence. Importantly, the peasant producers who occupied and tilled the land were generally not its owners: agrarian property was almost exclusively in the possession of a ruling caste of spiritual and temporal overlords.

For approximately four centuries following the Norman Invasion (1066), a feudal mode of production remained dominant throughout England. In Bloch's (1962: 244) opinion, the Normans' achievement was the establishment in England of a 'manorial regime of exceptional rigour'. Although the manor tended to develop along a typically feudal line in central England and in the southern Midlands, its purchase on other (especially marginal) regions was of a more limited kind. In certain areas, for example, peasants were more likely to be freeholders (as opposed to servile tenants) of estate lands, delivering money rents (rather than labour services) to the local lord. Anderson (1974a: 154) makes the point that, as a concrete social formation, England in the middle ages was, in fact, a composite social system 'in which other modes of production survived and intertwined with feudalism proper'.[1] Although most of the subordinate class of feudal England lived in a state of vassalage, some were slaves, while others remained free of any legal subjection.

None the less, most observers agree that the manor, in one form or another, was ubiquitous (Postan, 1972). It was within this socio-spatial setting that the vast majority of medieval countryfolk lived, usually in small group settlements, such as villages and hamlets. Although certain political, jural and economic features of feudal land organisation varied in space, the daily lives of the peasantry were marked by a similarity arising from the presence of common material conditions – such as the need for self-sufficiency, the difficulty of movement, the need to produce a defined surplus as rent and peasant control of the means of production. This similarity was further reinforced by the general absence of markets and commodity production.

Until now my comments have focused almost exclusively upon *rural* feudal society, the setting for most peasants' lives. Although urban areas were certainly a critical component of the medieval landscape, they were greatly overshadowed in material terms by the importance of the countryside. The overwhelming significance of rural space is apparent in terms of both demography and economic production. Hilton (1985: 121) reports that, in feudal Europe, 'the vast majority of the population – 80–90 per cent – was engaged in arable or pastoral farming'. Le Goff (1988) emphasises that it was the manor, or *seigneur*, which remained the centre of feudal production during the middle ages. Despite enjoying a progressive growth in both size and economic freedom throughout the period, the medieval town, none the less, remained dependent upon the seigneurial sphere for its basic material needs. In feudal England, the dominance of countryside over town was more marked than on the Continent. Anderson (1974a: 161) reports that in medieval England, 'Towns of any size were few and enjoyed no substantive independence'. England during the middle ages was devoid of the politically autonomous communes which developed elsewhere in Western Europe (Anderson, 1974a).

I certainly do not mean to trivialise here the significance of the rural–urban nexus in feudal society. Towns and cities played critically important roles as incubators of commodity relations that eventually spread into the countryside. Marx and Engels (1979) pointed to the separation of town (manufacture) and country (agriculture) as the first great division of labour. In their view, this socio-spatial contradiction was to become a driving force in the transition from feudalism to capitalism (Neale, 1975). With the Industrial Revolution, the setting of everyday life for the subaltern orders shifted rapidly towards towns and cities. Accordingly, the next chapter's enquiry into disability in the time of industrial capitalism will take the city as its principal setting.

Having sketched the general material context of feudalism – in particular, medieval England – I want now to shift the analysis to the two principal socio-spatial settings within which peasant communities were situated – the manor and the village.

The manor

The archetypal manor was a large estate of mostly agricultural land owned by a powerful authority (either an individual noble or a religious foundation), and worked by servile peasants. Each manor was divided into three parts: the demesne, which was the land directly 'worked' by the lord (with servile labour); the tenements, which were the small or medium-sized holdings of peasants; and the communal lands, such as pastures, meadows and forests. Manorial production was mainly agricultural, supplemented by limited pastoral activities.

The demesne was directly organised by the lord's stewards (reeves and bailiffs) and tilled by his or her villeins. The tenements were divided into units of land known as virgates (normally 30 acres), which the lord permitted peasants to use in return for labour services (usually performed on the demesne), or, in some cases, material rents of produce or money. The common holding for an individual family was one virgate, though many held less than this, and some held more (Hanawalt, 1986). Although ownership of the tenements rested finally in the hands of the ruling lord, virgates, or part-virgates, were usually held as customary tenures which could be inherited by succeeding generations of peasants.

In addition to servile tenants, many manors contained a small community of freeholders who enjoyed a number of legal liberties which placed them outside formal serfdom. These peasants usually rented or owned land within the manor, working it just as unfree tenants worked theirs. Practically speaking, the autonomy of freeholders was more often abstract than real, with their daily lives often differing little from those of their bonded neighbours.

The lord's income was principally derived from the cultivation of the demesne, the entire product of which she or he retained, with peasant virgates supplying a complementary surplus. In addition, the landlord's wealth was supplemented by a range of exactions on the manor's villeins; these were fees and fines for commonplace activities – ranging from illicit defloration (*legerwite*) to the brewing of ale – which were enforced by a manorial court (Duby, 1968).

Both commodity production and capitalist accumulation were all but absent from the manorial economy. Ruling lords were concerned simply with the exaction of maximum profit: 'The idea of reinvesting profit for the purpose of increasing production seems to have been present in few minds if any' (Hilton, 1975: 213). The manorial economy was also characterised by a weak, and often sporadic, circulation of money (Kosminsky, 1956; Duby, 1968).

As explained previously, a principal axiom of serfdom was the restriction placed on a peasant's mobility. Lords exercised control over where each of their servile tenants lived and travelled. These legal constraints on movement meant that the manor represented a powerful, though not impervious, set of boundaries around the social space of the peasant. Braudel (1981) has emphasised the manor as the spatially small world of the peasant; its border the outer limits of everyday life. Herlihy explains that 'The manor was a tightly disciplined community of peasants, under the rule and authority of a lord or *seigneur*' (1968: 3) (original italics).

Le Goff offers a portrait of this manorial world:

> The lord and the peasant found their needs satisfied in the framework of the manor, and in the case of the peasant, above all, in the compass of his home. Food was produced from the garden attached to the house and from the part of the yield from his smallholding which remained to him after he had paid his dues to the lord and the tithe owing to the church; clothes were made by women at home, and the basic tools – the quern or handmill, the distaff, and the loom – belonged to the family.
>
> (1988: 247)

With the exception, perhaps, of the occasional trip to a market town or religious site, many peasant lives must have been lived almost totally within the bounds of manorial estates. The manor was the centre of a set of dense, centripetal social relations for the peasantry. Legal proscriptions were more important to defining a peasant's personal mobility than physical capacities and limitations. For this reason, many physically impaired peasants must have maintained spheres of daily interaction similar to their neighbours.

The village

The vast majority of feudal peasants lived in small settlements – villages and hamlets, often with no more than 300 to 400 inhabitants. The village was the centre of peasant life; it was, as Gies and Gies (1990: 7) would have it, 'an integrated whole, a permanent community organized for agricultural production'. But an individual village was not necessarily the sole settlement axis of a single manor. Frequently, manor and village did not coincide (though they occasionally did); feudal estates were often linked to several villages, and *vice versa* (Kosminsky, 1956; Hanawalt, 1986). The inhabitants of any village might be distinguished by different manorial ties. In addition, variegation of the community arose from the division of labour: not all villagers were continually involved in agricultural pursuits, with many supplementing their income through the provision of artisanal services and unskilled labour to their neighbours.

The most powerful source of social stratification within the village, however, stemmed from the amount of land held by each peasant household, either as freeholders or serfs. Land was the principal means of production, and, accordingly, the extent of its ownership determined a peasant's standard of living. Most commentators (e.g., Postan, 1966, 1972; Hilton, 1975; Hanawalt, 1986) see the typical feudal village community as divisible into three general wealth strata of rich, middle and poor peasants. Kosminsky believes that most peasants can be allocated to the meso wealth level: 'The main body of the English peasantry, the villeins occupying virgates and half-virgates, were not rich, solid peasants, but a middle peasantry crushed by feudal exploitation' (1956: 240). Counterbalancing the effect of these social fault lines were the powerful forces which ensured a strong collective character for the peasant village. Hilton (1975) highlights the fundamental material imperatives – such as the need for villagers to co-operate with each other over pastures and harvests – which secured a high level of social cohesion among peasant communities.

For Hanawalt (1986), the typical village was marked by a concentric geography, with three main regions: the centre might contain a cluster of peasant houses, the church, outbuildings and gardens; the middle areas were the fields and meadows (both virgates and demesne); while the perimeter was often a rough area of woods and wastes. The whole terrain was known formally as the 'greensward', the boundaries of which villagers marked by an annual perambulation, or beating of the bounds. (Hanawalt, 1986). Even given that villagers might be bound to different manors of varying sizes, Hanawalt (1986: 21) believes that, for the typical peasant, a 'daily round of interaction was within a radius of five miles from the village'. Ault, similarly, emphasises the spatially-confined world of the peasant, observing that the English feudal village 'was "a world of neighbours" ... all ... within easy walking distance of each other' (1972: 15).

Having elaborated a general architecture of daily life in feudal England, I want now to address the specific question of how impairment was experienced within that setting. The revised typology presented in the previous chapter (Figure 4.3) suggested that three key sites framed the quotidian experience of impaired peasants (home, workplace and institution), each of which may be situated within the overlapping settings of village and manor.

Of course, such a conceptual-empirical frame in no way exhausts the range of social spatial experiences of impaired people. For example, scholars such as Hanawalt (1986) and Le Goff (1988) would doubtless argue that my typology overlooks other important village activity sites. Church, tavern, manor house and green, for example, were all important activity nodes in the feudal village which will, none the less, not feature in any significant sense in the following analysis. My concern here is to describe key patterns and relationships, not to describe the universe of phenomena that comprised the social space of disability in feudalism. Thus, the typology focuses the analytical 'gaze' on key, interrelated nodes of experience.

Impairment and everyday life

The home-workplace in feudal England

The spatial dimensions of home and work

Work-space and the domestic sphere closely overlapped for the peasant household (Mackenzie and Rose, 1983). The family cottage was the pivot of both production and reproduction activities. Due to the relative absence of wage labour in the peasant economy, there was no distinction between paid and unpaid labour within the peasant household economy. The domestic sphere was also a work site in that it was there that the primary products of peasant virgates and common lands were transformed into use values. According to Nicholson, 'Home was a workplace; raw commodities such as grain, milk, skin and wool were transformed there into the necessities of life' (1988: 33).

But as this suggests, home and workplace were not completely co-terminous for peasants: several important work sites – the family virgates, the lord's demesne, the common lands – lay beyond the confines of the family cottage. However, as Hanawalt (1986) has observed, these external places of work tended to be located close by the family home, with the most distant being no more than a few miles away. The routine of daily labour rarely took villagers beyond these external work sites, though some would have made an occasional trip to a nearby village or market town as part of their labours.

The land immediately surrounding the family cottage was known as the croft. The croft was delimited by walls or ditches and usually contained a house, the family garden, a barn, and, perhaps, other outbuildings. Croft sizes varied considerably, of course, depending upon the wealth of individual families.

While everyday labour was not distinguished by different forms of reward – that is to say, by the payment of some work activities and not others – it was certainly gendered (Hilton, 1975; Labarge, 1986). Men were mainly, though not exclusively, involved in agricultural pursuits, while women's work was chiefly, though again not only, child rearing and the production of household necessities

(Middleton, 1988). These sex-based allocations of work were pervasive, but not immutable; all household members, for example, were expected to help with major tasks, such as crop harvesting. Segalen (1983) has emphasised the fact that peasant women and men often co-operated in the execution of a range of tasks, including ploughing, sowing, harvesting and the processing of primary products (e.g., corn stripping).

This gendered division of labour corresponded to a bifurcation of work-spaces for women and men. Hanawalt's extensive study of medieval coroners' rolls provides empirical confirmation of the gendered work-space of the peasant household:

> Compared to men, women's accidents indicate that they spent much more of their workday around the house and village ... The place of death, there-fore, confirms women's chief sphere of work as the home and the men's as the fields and forests.
>
> (1986: 145)

Again, one must heed Segalen's (1983) caution against regarding this spatial division too schematically, as the spheres of men's and women's work frequently overlapped. Home and field should be considered as the centres of women's and men's labour respectively, rather than as the exclusive boundaries of gendered work territories.

The external context of the peasant household

The peasant household was little concerned with markets or commodity rela-tions. Commodity relations in the middle ages were chiefly confined to larger towns and cities, with rural production being mostly predicated upon the local-ised needs and conditions prevailing in village and seigneurial communities (Blaut, 1976). Le Goff (1988: 222) believes that the 'aim of the medieval economy was subsistence', rather than accumulation. Medieval theology, in fact, proscribed the accumulation of wealth and championed the peasant family occupying *terra unius familiae* – a portion of land that could support an average household – as the ideal social unit for the lower orders (Le Goff, 1988).

Hilton (1985: 5) also points to the self-sufficiency of peasant families, stressing that 'most of their economic production was for self-subsistence and economic reproduction'. Peasants, the direct producers of the middle ages, controlled (if they did not own) the means of production – principally, land and animals. Yet, while the peasant household was mostly self-sufficient, it was not completely closed to external imperatives. The most salient external constraint on the peasant house-hold was obviously the need to meet its obligations to its landlord. The most common form of exaction was labour service, with villeins required to spend as many as three days a week (though this varied from manor to manor) working directly for the lord on the demesne, or in the manor house (Lucas, 1983). In addition, the peasant household was subject to various other external compulsions in the form of tithes, money rents, fees and fines to the lord, ceremonial expenses, and the need to defer a portion of consumption for inheritances.

The peasant economy was largely cashless, though not completely so. Although local in range, and marked by low transaction velocities, markets did exist within and between villages, providing an opportunity for peasant households to raise limited amounts of cash for specific needs. Hilton (1985) notes that most peasant families tended to market a small portion of their product in order to raise funds for rents, taxes and fines. None the less, he stresses that:

> We must not ... imagine that these were small-scale capitalist farmers. Very little cash was retained after the payment of dues, and inputs of labour and materials were largely provided from within the family economy.
>
> (p. 129)

In addition to selling a small portion of its produce, a family might also supplement its cash income through the sale of labour (mostly to wealthier neighbours) and artisanal services (many peasants doubled as carpenters, blacksmiths, tilers, leatherworkers and the like).

Women produced in the home much of what peasant households sold for cash, or exchanged for other goods. The products of these cottage industries were, in effect, use values made by women for their own household's consumption, only produced in greater quantities. Cottage goods made and sold by women included ale, bread, butchered meat and cloth (Hanawalt, 1986; Labarge, 1986). These supplemental economic activities of women were critical in deciding whether a household was able to meet its external obligations without compromising its subsistence level (Labarge, 1986). Women also performed labour services in the home by weaving flax and wool for the local landlord (Duby, 1968).

In summary it may be said that the peasant household was a largely self-sufficient economic unit which had to satisfy certain clearly enunciated demands imposed upon it by the ruling classes. The most important of these obligations were the exactions through which the non-producing land-owning class confiscated the surplus product of the peasantry. The only other significant extraneous control on the household which can be named was the expectation that peasants would, in the course of their labour, follow certain religious and cultural traditions, such as the observance of holy days.

The two central economic imperatives facing the peasant household were first, the need to meet external charges on its product, and second, the necessity of maintaining its own subsistence. It is important to note that peasant producers were largely free to determine their standard of living; the only immutables (beyond the external compulsions) were the need to maintain adequate shelter and a minimum calorific intake for family members. Moreover, the peasant family was at relative liberty to decide the *manner* in which it satisfied both its internal (reproductive) needs and external obligations. In everyday life, a great diversity of responses would have been made by individual households to these twin needs.

The peasant household's self-sufficiency was consonant with a significant power for self-determination both in the level of its production and in the means by which this was attained. Families could exercise significant self-determination regarding both the extent of their needs and the ways in which these were gratified. The relative autonomy of each family in designing and executing its domestic economy is of critical interest to this analysis. In particular, the flexibility of the household in determining the form of its labour process implies much about the potential situation of physically impaired peasants, and it is to this issue that I now turn.

General features of the household labour process

In the following comments I oppose the frequent assumption of medievalists that the peasant household labour force was composed simply of all 'able-bodied' family members. Historians such as Hanawalt (1986) and Labarge (1986) correctly state that the subsistence nature of the peasant economy meant that households relied on productive contributions from *all* members in order to survive. But Labarge (1986: 163) is surely wrong in insisting that peasant work was 'labour involving all *able-bodied* members of the family' (emphasis added). Labarge here is clearly transferring modern conceptions of the 'able body' to her analysis of the peasant household. This is a questionable course, as it forecloses on the real possibility that very different ideas of physical capability prevailed in previous historical eras.

A central predicate of the peasant economy in any context is the need to balance the number of mouths fed with the number of hands deployed in productive activity. This requirement was realised in English manorial custom which forbade exemption from compulsory labour services on the basis of sickness (Duby, 1968). One might say that the imperative for universal work meant that peasant households *could not afford to consider any bodies as unproductive*, and that suitable types of work had to be found for all family members. As evidence for this, Hanawalt reports that manorial records,

> show that the aged remained as physically active and as involved in daily work as they were able. Even an old blind woman could be pressed into baby-sitting during harvest.
>
> (1986: 237)

Indeed, I shall argue that the material context of feudal production allowed peasant households a great degree of liberty in designing everyday tasks that would match the corporeal capacities of each family member.

Given its division on gender lines, the labour process of the peasant household should be analysed separately for women and men. It has been noted that the social space of peasant labour comprised two closely overlapping domains – the croft-centred activities of women and the field-work of men (Figure 5.1). However, before beginning this two-stage component analysis of peasant labour, a number of qualities common to both work spheres need to be pointed to.

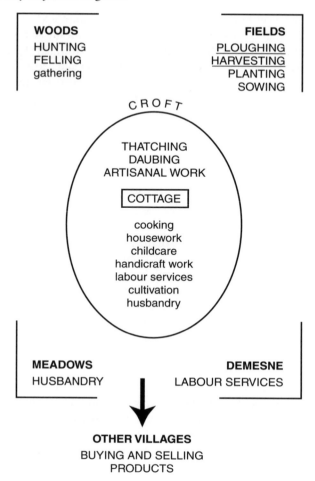

Figure 5.1 gendered realm of peasant labour content:

WOODS
HUNTING
FELLING
gathering

FIELDS
PLOUGHING
HARVESTING
PLANTING
SOWING

C R O F T

THATCHING
DAUBING
ARTISANAL WORK

COTTAGE

cooking
housework
childcare
handicraft work
labour services
cultivation
husbandry

MEADOWS
HUSBANDRY

DEMESNE
LABOUR SERVICES

OTHER VILLAGES
BUYING AND SELLING
PRODUCTS

Note:
Tasks undertaken by men denoted in UPPER CASE
Tasks undertaken by women denoted in lower case
Tasks undertaken by both sexes in UPPERCASE, UNDERLINED

Figure 5.1 The gendered realm of peasant labour

The first general consideration is the profoundly different manner in which time was reckoned in feudalism in contrast to the temporality of the capitalist era. In a celebrated essay, Thompson (1974) explained the pre-industrial understanding of time as dependent upon 'natural' (i.e., ecologically sourced) rhythms, rather than mechanical processes. Peasants conceived time through the passage of the seasons, the passing of the day, and the growth of crops and animals (Le Goff, 1988). Humanity itself was part of this 'natural time', with fecundity and birth providing important rhythmic dimensions to the peasant sense of temporality.

This is to say that the temporality of everyday life emerged partly from the individual's encounter with her or his immediate organic surroundings. Time for the average peasant was an organising principle revealed by the seasons (and, for that, self-evident): she or he had no need of any other external reference piece, such as a clock (Thrift, 1990). The limited number of time measurement devices available – sundials, hourglasses, and candles – were imprecise, and themselves open to the vagaries of nature (Le Goff, 1988).

One formal, inorganic constraint on medieval temporality was the liturgical calendar of the Church. In tracing the story of Christ, from Advent to Pentecost, the calendar prescribed the observance of various feasts and holy days, along with other religious practices. Even here, however, the patterning of observances was loose, and, to some extent, determined by the seasons (Le Goff, 1988). The effect of this organically sourced temporality was to make 'The organisation of the day, week and year in rural areas in Mediaeval times ... rhythmic rather than measured' (Thrift, 1981: 58).

Broadly speaking, the working day was framed by the general constraints of seasons, religious observances and certain communal laws, such as the common prohibition on night work (Langenfelt, 1954; Mandel, 1968). The effect of these constraints was that the duration and number of days worked in the middle ages were shorter and fewer than that demanded of the rural and industrial pro-letariats in later centuries (Pahl, 1988). In support of this, Mandel (1968) re-ports the work of Espinas, who estimated the average number of working days in the medieval era to be 240 in a year.

But within this temporal framework, the reckoning of peasant labour time was, as Thompson (1974) points out, *task oriented*: people simply worked until the job required (which they, in any case, determined) was done. Thus, labour time for peasants was a product of the combined influences of immediate organic imperatives and the task at hand. It is important to remember that the peasant's *body* was itself a powerful organic source of temporality; no doubt the extent of tasks, and the duration of labour time, were influenced by this. The power for self-determination of tasks meant that individuals could match work rhythms with their corporeal abilities.

Peasant labour time was thus discontinuous and open to sudden changes in extent. Work and social intercourse were closely interwoven, providing a further source of sporadic interruptions and irregularity of duration in labour. There can be no doubt that such reckoning of labour time would have permitted peasants to tailor their work efforts to fit their individual bodily capacities and limitations. Feudal temporality was a significant contributor to 'somatic flexibility' in the peasant labour process.

The feudal labour process was embedded within peasant social relations marked by cohesiveness and mutual support. Hanawalt (1986) avoids sentimentalising this feature of peasant life (a failing common to many romanticised views of the middle ages), viewing the social cohesion of the peasant community as a straightforward material necessity which ensured its survival. Co-operation and mutual dependency were critical in a society so often precariously balanced in terms of material resources.

The first, and most important, social ligature was the kinship bond. Again, the temptation to romanticise peasant society – in this case as a cohesive network of extended families – must be rejected in favour of a more complex truth. Both Segalen (1983) and Hanawalt (1986) found that the nuclear family was the most common feudal household form, thus contradicting the widespread belief of modern social scientists in the extended family as the predominant pre-industrial household form. The villagers in Hanawalt's (1986) study were more likely to rely closely on neighbours than on extended kin. The extended family residing under a single roof was, however, not uncommon, with its most frequent form being a nuclear family living with grandparents.

The peasant social order was based on an implicit governing assumption that all had to contribute towards its material sustenance (Kumar, 1988; Malcomson, 1988). In certain cases designated material realms within the labour process were set aside for particular types of producers. Ault (1965, 1972) has reported that local agrarian by-laws in the middle ages specifically reserved the task of gleaning (gathering grain left over from harvests) for physically impaired members of the village community. This author estimates that a gleaner might have earned as much in a day as a reaper (Ault, 1972).

Another important general practice in peasant production was the sub-letting of land by those incapable of working it. Again, peasant communal structures worked to ensure both that valuable land was not taken out of production and that villagers were materially supported. Such arrangements were most commonly made by widows who would sub-let their property to a neighbour in return for some form of sustenance (Hilton, 1975; Labarge, 1986). Postan (1966: 626) found evidence of this practice involving people 'who found themselves unable to cultivate their land – widows, invalids, old folk'. Hanawalt details a particular case in an English village where 'a kinsman of Ralph Beamonds took over the tenement because Ralph was impotent' (1986: 230). She maintains that in certain instances where a tenant had become too 'impotent' to work a holding, community leaders would meet in the manorial court and arrange for a neighbour to take over the land in exchange for maintaining the person (Hanawalt, 1986).

Men's work

Peasant men were mainly engaged in agriculture and animal husbandry. Men's daily labours took them beyond the family cottage to their own fields, the village commons, nearby woods and the lord's demesne (Figure 5.1). Occasionally, male family members might travel to a nearby town or village to market a portion of the household's product. The social space of men's labour was thus somewhat more extensive than was the work sphere of women.

Peasant males undertook a great variety of tasks in the fields and meadows surrounding the village. The type of work carried out at any time was largely dependent upon the season, and included activities as diverse as ploughing, shearing, planting, sowing, felling trees, and thatching and daubing buildings. Harvesting involved the entire village community. In addition to agriculture and husbandry, men often engaged in part-time artisanal work around the family croft, acting variously as bakers, millers, carpenters and blacksmiths to the village. These jobs might particularly have consumed men's energies during the months of January and December when agricultural work was suspended. (Hanawalt (1986) suggests that most men actually spent this time relaxing by the hearth; women, no doubt, continued to work.)

Both agriculture and husbandry necessitated a diverse mixture of activities, each requiring different combinations of physical strength and dexterity. It is conceivable that certain physical impairments – i.e., those not resulting in a significant immobilisation of the individual – would not have prevented peasants from participating in these outdoor labours. The same may be said of the croft-based artisanal work of men. In both cases the peasant labourer was free to decide how each labour was carried out; that is to say, the amount of energy applied, the spacing and duration of breaks, the length of the working day and the level of productivity.

While there was a minimum set of tasks to be done in accordance with seasonal imperatives, male peasants, none the less, had significant autonomy in creating a labour process which suited their household's needs and abilities. In this very real, material sense, one may speak of a *somatic flexibility* in the work regime of male peasants.

Women's work

The labour of women was relatively, though by no means wholly, sedentary in character, most of it occurring around the family croft (Figure 5.1). Women's work was chiefly the reproduction of family labour power and the production of use values for the household's direct consumption needs. Women were engaged daily in a diversity of tasks, ranging from food preparation and child supervision to the production of household consumables, such as food, beverages and clothes (some of these produced for sale) (Gies and Gies, 1990). Female family members also contributed to the home economy through the cultivation of a croft fruit and vegetable garden, and by keeping a small stock of domestic animals, including, perhaps, a cow, a pig and various poultry (Labarge, 1986).

Female labour was not exclusively home-based: women ranged through woods and along roadways picking nuts, wild fruits, herbs and greens (those near seashores gathered shellfish). In addition, women helped from late July to early September with harvest work in the fields. Women certainly worked as hard as men, if not harder. Labarge writes: 'Peasant wives were full-time workers whose tasks were essential to their household's subsistence and comfort' (1986: 161). While being no less physically demanding than men's work, women's labour was probably more fragmented and composed of a greater array of tasks. Certainly, women's work required less mobility on the part of the individual. The task regime of women was, thus, a heterogeneous set of physical demands, ranging from relatively light labours, such as minding domestic animals, to those requiring far more effort, like butter and cheese making or beating flax. The regimen of jobs undertaken by women was open to a large degree of self-determination by individuals. The length of the working day, the duration and timing of breaks, the amount of energy applied – all could be set by the woman engaging in domestic work.

Given the above considerations, there is no reason to believe that a physically impaired person (of either sex) could not have been integrated into the peasant domestic economy. The flexible, sedentary regime of women's work could have been moulded to match an inestimable array of physical abilities. The modern mind must recall the very different reciprocities between particular labours and their rewards, or social significances, which prevailed in the middle ages. The relatively simple and physically undemanding jobs of lighting the household fire, and caring for the family's domestic animals, had an importance that is hard to appreciate today. Remembering the imperative to balance the number of mouths fed with a sufficient amount of productive labour, one can imagine that peasant households made use of the flexible domestic labour regime to ensure that physically impaired family members had meaningful and productive work.

With this review of peasant labour complete, I want to now examine the Norwich (1570) and Salisbury (1635) surveys of the poor which were referred to in the chapter introduction (see also Appendix). What follows now are reviews of these two data sets, beginning with the earlier census. My intention in examining both data sets is to locate the position of physically impaired persons within the enumerated poor of both early-modern settings.

Physical impairment in Norwich, 1570

The Norwich census of the poor identified some 2,359 people – about one-quarter of the city's population in 1570 – as poor. This is not, however, a survey of the destitute. Most of those enumerated (66 per cent of males and 85 per cent of females) were in some form of employment, and the census must be regarded as having captured a broad section of the city's lower class.[2]

My examination of these data identified a total of forty-seven physically impaired persons, dwelling in forty-six households. There were twenty-nine females and eighteen males among this number. Physically impaired people comprised approximately 2 per cent of the total enumerated poor. People with physical impairments tended to be older than the general population surveyed (Table 5.1).

More than half of the physically impaired poor of Norwich were described as working at the time of the census. Of these, a majority were recorded as living either with family or in a household (Table 5.2). An even greater proportion of non-working physically impaired persons lived with families or in households.

While caution is advised in dealing with such a small sample, it is interesting that a significant number of non-working physically impaired persons living in households were either very young or very old. A 'normal' life course of work might be suggested here: the elderly might have retired from previous labours, and the young may have later gone on to some form of employment. In addition,

Table 5.1 Age structures of impaired and total poor

	Impaired poor		Total poor	
	No.	%	No.	%
Age				
< 16	3	6	926	39
16–59	23	49	1,036	44
> 59	21	45	330	14
unspecified	1	–	67	3
Total	47	100	2,359	100

Source: Derived from data in Pound (1971)

Table 5.2 Employment status and living circumstances of physically impaired persons

	No. living with family or in a household	No. living alone
Working	14	10
Not working	19	4
Total	33	14

Source: Derived from data in Pound (1971)

several of the unemployed physically impaired persons are described as having a trade, or occupation, although it is impossible to tell from the text at what stage in their life they might have obtained their skills. These factors together suggest that a cursory reading of the census data might understate the labouring potential (both latent and expended) of the physically impaired persons enumerated.

Of those physically impaired persons described as working, the great majority were women, all of whom were engaged in domestically based production (Table 5.3). All but one of the women were involved in the manufacture and/or repair of textiles or garments. Of the men, three are described as working at pipe filling. Interestingly, within the overall population enumerated, pipe filling is predominantly a woman's task with females comprising sixteen of the twenty persons so employed. Of the four male pipe-fillers, three are physically impaired (and all are described as 'lame'). Thus it appears that the normally female job of pipe filling was undertaken by certain physically impaired males, presumably because it was a relatively sedentary task.

Most of the working physically impaired poor were female. This may suggest that the nature of women's labour in late Tudor Norwich was such that it did not preclude involvement by those with physical impairments. Of all the female poor enumerated, over two-thirds were engaged in spinning for the local textile industry (Pound, 1971). More than half of the physically impaired women surveyed were involved in either spinning or related activities. Presumably the home-based character of these pursuits allowed workers significant flexibility in deciding how their labour was performed.

Table 5.3 Occupation by gender of working physically impaired persons

Occupation	No.
Women	
Spinning	16
Knitting	1
Sewing	1
Distilling	1
Sub-total	19
Men	
Pipe filling	3
Labouring	1
Spit turning	1
Sub-total	5
Total	24

Source: Derived from data in Pound (1971)

Only five of the physically impaired males surveyed were working. A significant number of the unemployed physically impaired males were of working age. The reasons for this situation cannot be deduced from the census data. Perhaps the growth of commodity production in late Tudor Norwich had been most vigorous in the realm of men's work, with the result that, by 1570, the labour power of physically impaired males had devalued further than that of women. The absence of rural labour opportunities may have contributed further to narrowing the employment prospects for physically impaired males.

Presumably the domestic nature of pipe filling was conducive to participation by physically impaired persons. That physically impaired males did not involve themselves in other forms of domestic labour, such as spinning, is interesting. This is probably attributable to contemporary ideologies of gender which would have proscribed most of women's work for men.

Over half of the physically impaired poor of Norwich in 1570 were engaged in what appears to be meaningful economic activity. The age structure of those who were not working suggests that other physically impaired persons might have found employment either prior to, or in the years following, the census. Thus it can hardly be said that physically impaired persons were rigidly excluded from participation in material production in late Tudor Norwich.

Physical impairment in Salisbury, 1635

The town council of Salisbury conducted a survey of its poor in 1635. The previous comments concerning the rather loose conception of poverty in the Norwich census apply equally here. Like its Norwich equivalent, the Salisbury survey is best regarded as describing many of the city's less well-off citizens, rather than a marginalised and impoverished stratum.

A total of 108 households, containing a total of 249 people, were enumerated. For each household, the name, occupation (if any), and age of each member is listed. The amount of relief extended to each household is given, as are the weekly earnings of the employed. The physical impairments of the poor are noted.

A total of twenty physically impaired persons, living in twenty separate households, were identified. Nine of these people lived in Salisbury's Bedden Row poorhouse. All but one of the impaired residents of the poorhouse were aged over sixty years. Physically impaired people comprised approximately 8 per cent of the total population surveyed.

More than half (thirteen) of the physically impaired persons identified were employed and earning money. Unfortunately, the occupations of physically impaired persons are not given, and it is impossible to discern how the money was being earned. Many of the non-impaired poor were listed as being occupied in domestic tasks such as bonelacing, quilling, spinning and weaving. It is possible that the impaired poor earned their money through activities such as these.

Seven of the thirteen physically impaired persons for whom weekly earnings are listed were residents of the poorhouse. This is reflected in the age structure of working physically impaired persons (Table 5.4).

Table 5.4 Age structure of physically impaired persons by employment status

	Persons working	*Persons not working*	*Total*
Age			
< 16	1	–	1
16–59	4	2	6
> 59	8	5	13
Total	13	7	20

Source: Derived from data in Slack (1975)

The average weekly wage earned by physically impaired persons was substantially lower than the mean figure for the balance of the enumerated poor (Table 5.5). The range of earnings by physically impaired persons was also significantly smaller than that for the remainder of those surveyed.

Table 5.5 Average and range of weekly earnings for impaired and non-impaired poor

	Impaired	*Non-impaired*
Average weekly earnings	6.3d	11.3d
Range of weekly earnings	1.5d – 10d	2d – 3s

Source: Derived from data in Slack (1975)
Note: 1s = 12d

As with the Norwich survey, the temptation to 'over-analyse' such a small data set must be avoided. This caution is even more pertinent in the Salisbury case, though, in this example, physically impaired people comprised a greater proportion of the total population surveyed than they did in Norwich.

The Salisbury data again show a majority (65 per cent) of the physically impaired persons identified engaged in meaningful economic activity. The survey concerned itself only with those poor receiving parish relief, and it would be interesting to know how physically impaired persons among the non-supported subaltern class fared in the labour market. All of those physically impaired persons earning weekly income were in receipt of relief at the time of the survey. None the less, the data indicate the existence of a relatively non-exclusionary production process, allowing most physically impaired persons to contribute to their support through their own labour. Even those in the poorhouse were not totally reliant upon public charity, earning in some instances half, or more, of their weekly income through independent work.

I turn now to the third important material site in the feudal social space of impairment, the institution.

The institution in feudal England

There is no doubt that peasant communities assisted members who, for whatever reason, were unable to support themselves. Beier (1985) emphasises that manorial society bound and sustained even the poorest of peasants within local community settings until the end of the middle ages, and sometimes beyond. Early Saxon Poor Laws established that kin would domicile any peasant without shelter or means of support; after 1066, Norman governance intensified this form of bonding within manorial communities (Leonard, 1965).

It is a common misconception that feudal England was beset by roaming, rootless bands of impoverished peasants, drifting from village to village in quest of succour and, occasionally, theft and mayhem. The truth is that the social space of medieval England was relatively calm: the very real tide of mendicancy, to which this historical commonplace refers, did not in fact reach its high mark until the sixteenth century. The early modern era saw an explosive growth in the numbers of vagabonds and beggars at large in the realm. This phenomenon pressured the Tudor monarchs into passing anti-mendicancy laws of exceeding savagery. Beggars certainly roamed feudal England, but they were a relatively minor phenomenon: it was the pilgrim journeying to Westminster, Durham, or any of the other religious centres, who was the common wayfarer of the period (Clay, 1909).

My principal interest here is in the institutional resources of the local peasant community. Centuries before the 'Great Confinement' which Foucault (1979, 1988b) has so evocatively described, feudal England was spanned by an institutional network of considerable proportions. Feudal institutions varied considerably in size, function, location (rural/urban), and ownership, being variously, large monastic hospitals, 'houses of hospitality' for pilgrims, and small, general-purpose almshouses. Clay (1909) has estimated that there were upwards of 750 'hospitals' (a term she uses to describe the entire range of institutions just described) in medieval England. McIntosh (1991) has identified 978 residential institutions for the poor which operated at some time during a somewhat later period, 1388–1598.[3]

Although multifarious in character, medieval hospitals shared a quality which distinguished them from modern institutions established after the eighteenth century. The feudal institution was generally to be found in propinquity with, rather than excluded from, the community which it served. The one exception to this rule is the rather special case of the lazaretto, which was always located away from place of settlement (e.g., beyond a city's walls); feudal communities fearing the contagion of leprosy. Clay (1909) estimates that, of the 750 medieval hospitals, over 200 were given over to the care of lepers.

At this point I want to make an important conceptual-empirical qualification. Leprosy, both in its pathology and its social signification, represents a departure from the types of physical impairments being considered in this study. While I hope that geographers will in the future explore the specific historical experiences of people with leprosy, I will give it no further explicit attention here. Feudal communities were, themselves, well aware that leprosy was a necessarily contagious and progressive disease which distinguished it from other impairment-causing conditions. In view of the distinction given to leprosy by people in the middle ages, it is unfortunate that historians such as Le Goff (1988) tend to lump 'maimed' people together with lepers and the ill, thereby conflating the issues of sickness and impairment. I think it important to pause here for a moment to consider the implications of this confusion.

A royal edict of 1348 proclaimed that lepers were to be 'expelled from the communion of men' (Clay, 1909: 186). Village leaders were bound to expel the leper as an infectious danger to the community, and there is clear historical evidence that this occurred with some regularity. Though the occasional misdiagnosis might have occurred, such cases would have been rare: the clarity of leprosy symptoms, such as oozing sores, would have quickly set this disease apart from physical impairments. The leprosy generalisation in analyses of feudal social relations therefore encourages historical explanations of all 'cripples' as outcasts. This approach echoes the historicist shortcomings of disability analyses which were identified in the previous chapter.

It was common for a peasant village to contain a small almshouse, usually of no more than thirteen beds (Clay, 1909; Hanawalt, 1986). These modest institutions could be provided by a number of bodies, ranging from a religious fraternity to a craft guild. The squire of the manor would also, on occasion, maintain an almshouse for dependent tenants. In larger towns, a 'Maison Dieu' or 'bedehouse', as the almshouse was variously known, may have operated as an adjunct to a cathedral; in other cases it may have operated under the auspices of a corporate body, such as a municipality or board of trustees (usually made up of aristocratic and mercantile benefactors). Monasteries functioned as important, additional centres of relief for the needy (Pound, 1988). Abbeys such as Ely, Croyland and Glastonbury were famed for their generosity towards the poor (Leonard, 1965).

Although from a somewhat later era, the poorhouse in early Stuart Salisbury was essentially similar in scale and operation to those of the medieval period. At the time of the census (1635), the Bedden Row Poorhouse contained forty-six residents – nine of whom were physically impaired – reflecting the larger (though modest by later standards) scale of urban institutions. In this case the poorhouse was maintained by the town council (Slack, 1975).

Clay (1909: 15) believes that the 'majority of hospitals were for the support of infirm and aged people'. Just so: but it is difficult, as her analysis implicitly acknowledges through its generality, to specify a profile of the medieval institutional population (beyond identifying the significant presence of lepers). In the case of any village almshouse, one can imagine it helping to support a range of people who were unable to provide for themselves, including the sick, the elderly and the insane.

In the earlier discussion of the home-workplace context I demonstrated that physically impaired persons may well have made a valuable contribution to peasant household production. Even when isolated from a family unit, physically impaired persons may yet have sustained themselves through a combination of their own productive endeavours and various common strategies, such as subletting land to which they held rights. This argument refutes the claim that there is a necessary historical connection between physical impairment and social dependency.

Physically impaired peasants did end up as residents of village almshouses and larger hospitals – Clay (1909) mentions the presence of 'cripples' in several institutional cases – for all sorts of contingent reasons, not necessarily arising from disablement. The social dependency articulated through the village almshouse was frequently of a qualified kind. Village poorhouses were rather loosely run, with inmates free to carry on economic activities and contribute to their own support.

Some peasants no doubt avoided the almshouse through the protection afforded by guild membership. Many guilds, both urban and rural, provided disaster insurance which covered the loss of a limb or other impairments (Hanawalt, 1986). Compared with later social forms, peasant society presented fewer material impediments to an impaired person's chances of surviving independently. However, it was by no means ideal. The particular circumstances of some who experienced impairment at some point in their life (especially if suddenly) would have made special supports necessary.

For those needing recourse to the village almshouse for shelter, institutionalisation certainly did not mean banishment. The almshouse remained very much part of village social space, evidenced both by its physical proximity and by its residents' continuing ties to their peers. Village communities were too small and too cohesive (not always their most pleasant feature) to harbour institutions where villagers disappeared from everyday social intercourse; poorhouse residents must have remained as actors in the quotidian affairs of local communities.

Of course, reliance upon charity was not necessarily a social stigma in an era when influential Franciscan and Dominican thinkers extolled the spiritual virtues of poverty, and, even occasionally, physical frailty (Beier, 1985; Le Goff, 1988). Wealthy families were expected to contribute personally to the support of their poorer neighbours; the rich, indeed, sought opportunities for almsgiving, assured by Church teachings that benevolence maintained the 'health of the soul' (Rosenthal, 1972; Checkland and Checkland, 1974). At the local level, medieval philanthropy frequently provided partial supports for the needy, reducing the potential demand of the poor for recourse to the almshouse. McIntosh explains that,

> people of comfortable means provided assistance to their poorer neighbours in the form of food, clothing, bedding, the right to live in buildings at lowered or no rent, or nominal employment to the elderly or children.
>
> (1991: 2)

It is thus conceivable that village almshouses in the middle ages were rarely full, and even occasionally empty for extended periods. Although almost impossible to substantiate fully, some support for this claim is found in Laslett's (1971) work. In his study of 100 English, pre-industrial settlements, Laslett found that only 335 people, from a total population of around 70,000, were resident in an institution. On the basis of this, one is tempted to agree with Laslett who believes that the pre-industrial almshouse touched the lives of few peasants.

Conclusion: reflections on the feudal case

This chapter has explored the social space of impairment in feudal England through a two-stage analysis. The first part explained the boundaries of everyday life – and, hence, peasant social space – as essentially corresponding to the manor and the village. From this, the next part of the enquiry sought to explore the material context of impairment in these settings by focusing on the peasant household and the medieval institution. The analysis set the experience of impairment within the quotidian context of peasant life.

Le Goff (1988) believes that many feudal peasants were physically impaired. The middle ages, he says, 'were full of the maimed, hunchbacks, people with goitres, the lame and the paralysed' (Le Goff, 1988: 240). The evidence for this assertion includes the known prevalence of impairment-causing diseases such as tuberculosis. These diseases, combined with the generally poor peasant diet, and limited medical expertise, must have made impairment very common in the middle ages. Further support for the hypothesis that impairment was a feudal commonplace can be located in the ecclesiastical realm (Clay, 1909; Le Goff, 1988). The saints – those spiritual exemplars for, and of, the peasantry – included among their number a coterie associated with physical impairment (in England these were St Giles, St Godric and St Thomas). The numerous religious fraternities which provided support for cripples – some, like the Brothers of St Anthony, dedicated to this cause – are further evidence of impairment's prevalence.

Does this assumed ubiquity of impairment explain its relative invisibility in records surviving from the middle ages? I venture to say so. *Impairment itself* was probably a general feature of peasant social space in feudalism. Bodily impairment was doubtless an accepted, prosaic element of peasant life, and may only have marked itself out when, on occasion, it was seen to have spiritual significance; an example of this being the many miraculous cures of medieval cripples catalogued by Clay (1909). Additionally, certain types of impairment may have been the focus of community anxiety, or worse, when mistaken for evidence of leprosy. The physiological terrain was, as Le Goff (1988) relates, an uncertain object of socialisation, with extreme forms of embodiment taken variously to signify both good and evil incarnate. The ambivalent metaphysics of impairment in the middle ages suggests a corresponding ambiguity in its material context. Le Goff asserts that the medieval world was a fusion of spiritual and temporal worlds, a social space lived as a 'network of earthly and heavenly dependencies' (1988: 164). It is thus not unreasonable to suppose that this rather open spiritual treatment of impairment reflected something of its temporal socialisation.

Feudal society certainly did not dictate a single, ineluctable lived experience for physically impaired peasants. Indeed, a central precept of the embodied materialism that I outlined in Chapter 3 is the view that no society can totally impose a form of social experience on actors by reason of embodiment or any other trait. But the social constraints (and liberties) attached to certain forms of embodiment do vary between societies. I therefore want to conclude the chapter by reflecting upon the particular social relations which surrounded impairment in feudal England.

The material structure of peasant society certainly allowed physically impaired persons a significant degree of flexibility in determining their life course. Another way of saying this is to assert that the feudal social space of impairment was characterised by a structural pliancy which permitted – even encouraged – individuals to shape meaningful and productive lives for themselves. Critical to this analysis has been the demonstrated autonomy of peasant households in determining their own work regimes. It was shown that peasant families had the freedom to design labour processes which could match each household's physiological resources. Important here was the relative absence of external valuations on individual labour with households themselves retaining discretion over decisions about the work of members. Just as every family would have been distinguished by its physiological profile, so too would the labour regimes of each have taken individual forms; a common imperative for all, however, was the need for each household to deploy its entire somatic resources.

The argument is not that the social space of impairment was indistinguishable from that of the peasantry in general. Certainly, one may define a separate social space for impaired people insofar as people with this form of physical difference must have shared limitations on the sphere of everyday life. If the material context of peasant life was not ideal for the personal capacities of physically impaired persons, it was also not completely disabling. The type of work undertaken by men, for example, with particular somatic requirements such as strength and mobility, may have precluded involvement by some physically impaired persons. It is thus conceivable that physically impaired men frequently transgressed the gender division of labour and engaged in the relatively sedentary work of women.

There is no reason to suppose that this transgression was socially problematic, and there is limited empirical evidence of its occurrence in the Norwich survey.

A further characteristic distinguishing the sphere of impairment from the everyday terrain of peasant activities arises from the socially variegated experience of institutions. The preceding analysis has argued against the notion that physically impaired peasants necessarily, or even frequently, relied upon institutions for survival. However, it cannot be doubted that impaired persons did experience institutionalisation more commonly than did the general peasantry. An important consideration here is the social porosity of the village almshouse, residence in which hardly meant separation from the daily life of the village community.

Overall then, the social space of impairment must be seen as distinct from, *yet embedded within*, the general terrain of everyday life for the feudal peasantry. The domain of impairment may have differed from general social space in its physical extent, its gender contours, and the significance of its institutional outcrops, but the two terrains were not opposed to each other. The social space of impairment cannot be presented as marginal to the realm of everyday village and manorial life; it must, rather, be placed within the quotidian peasant landscape. Again, this is not to deny the singularity, or heterogeneity of forms, of everyday life for impaired peasants; this is simply to oppose the notions that these differences always either placed impaired people outside the congress of peasant life, or distinguished them as dependent and burdensome members of the community.

At the individual level, this investigation of the structures of peasant existence did not expose any material reason why impaired family members could not have remained *in situ*, contributing to their household's sustenance. Hanawalt's (1986) review of manorial court records and coroners' rolls has led her to reject specifically the suggestion that peasants practised infanticide on physically impaired infants. One imagines that as physically impaired children grew, they were encouraged, even harshly forced by modern standards, to define for themselves an everyday regimen which contributed to the life of their household.

As argued in Chapter 3, the social space of impairment in any society is best conceived as a bounded set of possibilities and restrictions, rather than an ensemble of predetermined compulsions, emerging from the material qualities of peasant life. What is at issue in different times and places is the relative strengths of constraints and opportunities for impaired people. In feudalism, the impaired peasant body was, to a significant degree, an autonomous creator of social space. Disablement, as the antithesis of this power for self-creation, was not an attribute of the material structures surrounding, and underpinning, peasant life. As Fraser (1997a) reminds us, feudal social relations were characterised by interdependency, unlike their capitalist successors which elevated individuality to the highest moral (and material) level. It was not shameful to be poor because the bourgeois notion of 'economic independence' was yet to be born. Within the complex, layered dependencies which constituted feudal village life, physically impaired people were not isolated as 'social dependants' – this abject identity was a construction of the capitalist social order.

This is not to indulge in the sort of misty-eyed nostalgia evident in some – especially, communitarian – assessments of feudal village life. In material terms, life was often hard for all peasants and the village was, as Harvey (1996) points out, an oppressive, otiose social organisation which many escaped at the first opportunity to do so. The point is that the particular social oppression of disability was weakly developed, even largely absent, within the feudal social form. As the following chapter will show, the rise of capitalism was to delimit this capacity for self-determination by bringing new, compulsive socialisations to bear on the body.

6 The social space of disability in the industrial city

Introduction

This chapter explores the social space of disability in the industrial capitalist city, focusing on one important example, colonial Melbourne.[1] Specifically, my aim is to situate the street life of disabled people within this unique historical social space.

Why focus on the street? As I will show, much of the surviving historical evidence about disabled people in industrialism locates them in street settings, usually as *displaced* figures marginalised both from formal public spaces and domestic realms. My analysis will explore the significance of 'the street' to disabled people among the proletarian and lumpenproletarian classes, while also shedding some light on the 'uneventful lives' which passed quietly within the institutional and domestic spaces of the industrial city. I shall broaden the picture's frame by drawing upon some English accounts of everyday streetlife in the industrial city. My examination will rely again on the concept of social space outlined in Chapter 4. However, I shall modify this conceptualisation by introducing a fourth node, the 'street', in order to reflect its importance to disabled people in the industrial city.

Why focus on colonial Melbourne? Two reasons: first, because I have considerable empirical familiarity with this historical setting (see Gleeson, 1993, 1995b), and second, because Melbourne is a worthy exemplar of the industrial city (Davison, 1978). It may help here to expand a little upon the latter ground. The capital of the Colony (now State) of Victoria, Melbourne had an 1891 population of nearly half a million. By the late nineteenth century Melbourne was regarded as one of the premier cities of the British Empire with a rateable value surpassed only by London and Glasgow (Briggs, 1968). By the early 1890s, the city's extensive manufacturing sector employed about 30 per cent of the male labour force. Most industrial establishments and the proletarian labour force were located in the inner ring of suburbs circling the Central Business District (CBD) (Lack, 1991). The fragmentary historical records of life in this industrial, proletarian core suggest the presence of a considerable, if marginalised, population of disabled people.

The chapter is in three parts. The first charts a broad account of the transition from feudalism to capitalism (focusing on England) which highlights the significance of this modal shift for physically impaired people among the lower social strata. The second part situates the phenomenon of disability in the industrial city. The final part of the chapter sketches the streetlife of disabled people in the industrial city, focusing on the case of colonial Melbourne.

The transition from feudalism to capitalism

Capitalism became the dominant mode of production in Europe only after a prolonged and geographically uneven process of development, whose origins can be traced back to the early stages of feudalism itself. In England, a series of shattering blows were delivered in the fourteenth century against the social space of feudalism in the form of wars, civil disturbances, famines and plagues (Genicot, 1966). These calamities combined with a rather more gradual, and uneven, process of economic change – essentially, the spread of markets – to effect a far-reaching transformation of the English political-economic landscape (Neale, 1975).

The modal shift from feudalism to capitalism was in fact the product of a series of more specific processes of change that affected the various structures of everyday life in different ways. Some shifts were sudden and widespread, such as the Dissolution of monasteries, hospitals and abbeys in the 1530s and 1540s, while others, such as the erosion of the manorial economy, were slower and followed irregular spatial patterns.

The decline of the feudal economy

In economic terms, the transition from feudalism to capitalism was a long and complex process, marked by key evolutionary stages in the development of markets, notably mercantilism and industrialism. The growth of markets and money from the twelfth century led to an expansion in petty commodity production among the peasantry. By the late seventeenth century, commodity relations had gained an enduring purchase on English social space. The growth of a money economy, together with the increasing size and range of markets, led to the peasant household redirecting an increasing portion of its energies towards the production of commodities, such as cloth, food and beverages. With this shift in its productive focus the peasant household conceded a significant degree of its autonomy. The household was no longer concerned solely with the production of its own material needs, and henceforth it was subjected to a powerful external force – the market – which defined the value not only of its product, but also of its labour.

The rise of commodity relations profoundly changed those processes of social embodiment that were sourced in work patterns. In particular, this political-economic shift lessened the ability of disabled people to make meaningful contributions to their families and households. Markets introduced into peasant households an abstract, social evaluation of work potential based upon the law of value; that is to say, the competition of labour-powers revealed as average socially necessary labour times. This productivity rule devalorised the work potential of anyone who could not produce at socially necessary rates. As households were progressively drawn into dependence upon the competitive sale of labour power, their ability to host 'slow' or dependent members was greatly reduced (Mandel, 1968).

Moreover, as capitalist relations strengthened and took root within a wider field of feudal society, the tendency was for the average socially necessary labour time of commodities to be driven down. Thus, one may envisage a historically iterative process where successive new averages are struck as the general conditions of productivity improve. Yet each historical 'round' occurs in a material context of socio-spatial evolution where both the labour process and the level of technological development are continually being remade. Shaped by the law of value, this material development assumed an implicit corporeal bias: henceforth space was manipulated in ways that ensured maximum productivity from those bodies valorised by the market. In short, the growth of markets progressively destroyed the socio-cultural contexts that had valorised the labour and social contributions of disabled people in the feudal era.

None the less, the erosion of the feudal peasant economy occurred over a long period and was fiercely contested by the lower orders themselves. By the early eighteenth century, the vast majority of England's people still lived in rural areas. Though capitalist relations now dominated the countryside, certain critical structures of everyday life had remained unchanged since the feudal epoch: the village continued to be the pivot of social space for peasants, many of whom still rarely ventured beyond the boundaries of their home community (Genicot, 1966); and the family cottage was still the centre of the family's production and reproduction activities (Malcomson, 1988). Wage labour was irregular and often in short supply, and families in the lower orders survived by combining occasional paid work with domestic industry (putting-out), artisanal work and, if they were lucky enough to retain some land, agriculture and animal husbandry (Laslett, 1971; Middleton, 1988).

As a spatially diffuse and domestic form of production, the putting-out system still permitted an appreciable degree of self-regulation by peasant workers. Merchants were able to pressure cottagers into working extended hours, but peasants none the less retained control over how this time was filled (Lazonick, 1990). Sharpe writes that 'long hours might be ameliorated by a degree of control on the part of the worker on how long was worked at one stretch, or how quickly a piece of work was completed' (1987: 207).

Through resort to domestic industry, many households would still have managed to configure their internal work regimes to match the diverse bodily capacities of members. Indeed, the autonomy preserved by putting out for the peasant producer was a source of irritation to merchants who complained of workers thieving from the raw materials advanced to them (Doray, 1988). The response of the merchant bourgeoisie was to manipulate the spatiality of production by starting to consolidate small groups of workers in particular sites to form manufacturing ensembles. The system of manufacture, in which workers became explicit wage earners (rather than being paid for output as domestic producers were), constituted a bridge to the large-scale factory which was to emerge following the Industrial Revolution.

By the eighteenth century, most rural lower-order families maintained an extensive, rather than intensive, economy (Malcomson, 1988). This meant that households knitted together a wide range of supporting activities in order to minimise family dependence upon wage labour, which remained sporadic and unreliable. Apart from paid work, families undertook handicraft activity, and, depending upon access to land, agriculture and husbandry in order to sustain themselves.

The long process of land enclosure made this struggle to remain autonomous of exchange relations increasingly difficult.[2] The growth of rural commodity markets (particularly for wool) made large-scale pastoralism attractive to landowners who proceeded to convert arable land into sheep pasturage. These consolidations were often achieved through forcible enclosures, frequently accompanied by the eviction of former tenants (Hilton, 1975). The loss of common rights to lands which had provided peasants with access to grazing, fuel and raw materials drove many who had escaped forcible eviction from their villages and estates. The combined effect of these dislocations was to create a landless social stratum that increasingly took the form of a proto-proletariat as its dependency on rural and urban wage labour intensified.

None the less, even by the mid-eighteenth century, exchange relations had not gained a universal purchase on the rural household sector, and families often continued to produce many of their own use values. Therefore, the situation of impaired peasants must have varied in relation to ability of their households to remain autonomous of commodity relations. No household, of course, could have completely sealed itself off from the importunate press of markets, but many may have greatly reduced their dependence upon exchange through domestic industry, private farming and husbandry, and the exercise of rights to common land.

The growth of rural capitalism certainly confronted impaired people with powerfully disabling forces. But the character of peasant social space – in particular, the ability of the lower orders to maintain their home-workplaces as redoubts against market relations – was to provide impaired people with some protection against the forces of social devalorisation. This sanctuary was not to last: the advent of industrial capitalism in the second half of the eighteenth century saw the final dissolution of the peasant landscape, and its replacement by a social space which admitted impaired people only as disabled dependants.

The rise of dependency

Socio-political relations evolved gradually, but relentlessly, towards the bourgeois democratic forms that were achieved in France and Britain during the late eighteenth and early nineteenth centuries. The social bonds of feudal society were progressively supplanted by new capitalist social relations founded on the ideal of self-reliant and free labour. Associated with this change were the new, official orthodoxies about poverty and economic dependence that served to reinforce the arrangement of labour through markets (Fraser, 1997a).

In England, from the middle of the fourteenth century, a succession of laws concerning the relief of poverty and the punishment of vagrancy were enacted by a state which had previously shown little inclination to legislate on either issue. The labour shortages and peasant unrest unleashed by the Black Death (1348–49) were doubtless responsible for the monarchy attempting to re-establish the compulsion to work by legislative means. Most measures – including the 1349 Statute of Labourers which aimed to stabilise commodity prices – were resisted by a peasant stratum agitated by the presence of increasingly virulent and disruptive market forces. By the fourteenth century, the corrosive influence of commodity relations was in evidence, progressively dissolving feudal social ties (notably, serfdom). This, combined with the Dissolution of the monasteries during the 1530s, reduced the capacity, or willingness, of local communities to support dependent members, thus swelling the ranks of the rootless, indigent poor throughout the country. Tudor governments during the sixteenth century were convinced that the realm had been overrun by hordes of roaming vagrants who resorted to crime, importunate begging and other outrages as a means of avoiding honest labour (Beier, 1983, 1985).

The last (1601) Tudor enactment on poor relief is testimony to a state by now committed to legally enforcing capitalist labour relations. The 1601 legislation, which was to survive essentially intact until its amendment in 1834, introduced a strict sense of dependency based upon a physical inability to labour, and established a system of compulsory local taxation (parish rates) to support the relief of the 'impotent'. Thus physically impaired people became established – in law at least – as social dependants, whose proper place in the new market order was on the economic margins reserved for those unable to sell their labour.

Conversely, the Act also contained separate provisions for the 'sturdy' poor clearly aimed at compelling them to remain in the social labour stock. The principal of these punitive measures was the prescribed establishment of Houses of Correction which were to punish the idle through harsh, confined labour. Elizabeth's law was only slowly, and irregularly, implemented at the local level, and by the early nineteenth century had become the object of extreme political anxiety for the new industrial bourgeoisie. By then it was claimed that the regulation of the poor had degenerated, indeed become perverted, such that the system of parish relief was actually encouraging, rather than preventing, economic dependency among the lower orders. Doubtless, the rather lax application of the poor laws in the eighteenth century benefited disabled people by diluting the implications of codified dependency.

However, this state of affairs was not to last – the 'democratisation of dependency' deeply alarmed the new industrial bourgeoisie and its leaders set out in the 1830s on a vigorous campaign to re-crystallise the independence–dependence dualism that had been established by the Tudor law. The result of their efforts was a new Poor Law Act in 1834 that laid out a strict distinction between the ablebodied poor, who would be compelled to labour, and the disabled poor, who were considered 'fit objects of charity'.

The rise of the industrial city

The Industrial Revolution

The term 'Industrial Revolution' has become a commonplace of history and is generally taken to mean the series of complex socio-economic transformations which began in the second half of the eighteenth century and which culminated with the eventual preponderance of the factory system over other productive forms some time in the late nineteenth century. The following summary depiction abstracts from this rich historical continuum of change by concentrating on the nineteenth century, during which England experienced rapid socio-spatial changes, in the form of urbanisation, colonial expansion, and the spread of the factory system.

By the late eighteenth century, centuries of growth by commodity relations had removed most of the major elements of feudalism from the English social terrain. The ground was now clear for the erection of the unique superstructure of industrial capitalism. While the architecture of the new social formation was to retain the familiar features of home, workplace and institution, their arrangement and function were to be radically altered by industrial capitalism. The new proletarian landscape was, of course, distinguished from its feudal and early modern predecessors by a socio-spatial division of paid work and reproduction (Berg, 1988). This feature of proletarian social space was to have a significant impact upon the everyday lives of impaired people.

The separation of paid work terrains from domestic space and the rise of the modern institution both played a critical role in the materialisation of patriarchal capitalist work relations by creating distinct, if often imbricated, social spaces for labour and non-labour. I argue that these enclosures of labour and non-labour in separate material domains were realised through three successive territorial confinements of social groups.

Three confinements

Foucault's (1979, 1988b) argument for a 'Great Confinement' of the poor in seventeenth-century Europe finds only weak support in English institutional history. Foucault's idea has more relevance to the English case if it is resituated within the historical context of emerging industrial capitalism. In the seventeenth century, the start of the first of three 'Great Confinements' was in evidence as the early industrial capitalists began to concentrate both workers and production in factories. This initial move to enclose labour was achieved through a general confinement of direct producers. Early manufactories often enlisted women, children, the sick, and impaired and old persons as workers (Mandel, 1968). Later, the first factories remained dependent upon a polyglot workforce of men, women and children – particularly the latter two groups, whom industrialists regarded as cheap and 'tractable' sources of labour power (Briggs, 1959; Hobsbawm, 1968). In the early stages of industrialism, non-labour had little social meaning with capitalist manufacturers super-exploiting a diverse workforce (with tacit state approval) as a means of overcoming the opposition of male craftworkers to the factory system.

Moreover, manufactories, and later factories, existed alongside a large domestic production sector. Indeed, domestic production remained the basis of the expanded textile industry until well into the nineteenth century (Lazonick, 1990). But, by the 1840s, factory production with the superior power-loom had eclipsed the domestic system, causing a precipitous decline in the latter (Ashton, 1948). Thus, a general enclosure of labour and the forces of production was not achieved until the latter stages of the Industrial Revolution.

The enclosure of non-labour began in earnest in the wake of the 1834 amendments to the Poor Law. This brutal law proclaimed that all 'objects of charity' be enclosed in a new national system of workhouses. Of this, Durkheim noted that, 'The insane and the sick of certain types, who were heretofore dispersed, [were] banded together from every province and every department into a single enclosure' (1964: 188). For Higgins, the workhouse was a 'pen of inutility':

> The workhouse, the true shrine of the work ethic, was a sort of concentration camp in which were incarcerated, and held up as an example, those who admitted their inutility to capital – the sick, the mad, the handicapped, the unemployed – and in conditions which were even more monstrous than in the factories.
>
> (1982: 202)

None the less, as Driver (1993) shows, the new law was far from monolithic in its effects, and workhouses were constructed and operated in a variety of ways at the local level. Some localities in Northern England, for example, resisted the new law, and there was considerable variation in the application of the legislation's more severe edicts. Thus, the enclosure of disabled labour power in workhouses was never universal and occurred through an uneven geographical process. The later establishment of hospitals and purpose-built institutions for disabled people by both public and charitable bodies would considerably extend the landscape of social dependency.

Since its first appearance, the bourgeois social form had been redefining labour relations, both by disabling impaired people and by devalorising the work of women. The general enclosure of labour within factories was followed in the late nineteenth century by a further enclosure as many forms of work became the exclusive preserve of non-disabled men, while women, children and impaired people increasingly shared a common social status as non-labour. One might say that within the non-labour category a distinction came to prevail in both state policy and social attitudes between 'incapable' (sick, impaired and old persons) and 'inappropriate' (women and children) labour powers.

By the late Victorian period, a new gendered division of labour between men as producers of value and women as reproducers of labour power had begun to gain widespread acceptance among both the bourgeoisie and the male proletariat

(Hartmann, 1979; Mackenzie and Rose, 1983). None the less, while child labour declined in the second half of the nineteenth century, women still remained heavily involved in a variety of industrial occupations until much later in spite of both increasingly prohibitive factory legislation and the opposition of male craft unions to competition from lower-paid female workers (John, 1986).

The three social confinements just outlined shaped the proletarian terrain which emerged in the second half of the nineteenth century. This new social space differed markedly from that of the feudal peasantry by introducing sharp functional distinctions between home, workplace and institution in everyday life. These confinements also realised a marked distinction between labour and non-labour in social space by reserving the new industrial workplaces for the former and by reconstituting domestic and institutional domains for the latter.

The disabling city

The production of deviant bodies

The principal (paid) productive loci of industrial England in the late nineteenth century were factory, mine, forge, shipyard and railway. These new enclosures of labour power were achieved through a demographic concentration evidenced in the spectacular urbanisation of England during the nineteenth century. The first stage of industrialism had spawned new cities, sometimes conceived from existing towns (Manchester, for example, grew from 17,000 inhabitants in 1760 to a population of 303,000 in 1850), mostly in textile-dominated Lancashire (Wohl, 1983). Established cities also continued to expand, and, by 1851, more English people lived in urban areas than in the countryside, with almost one-third of Britons residing in towns of over 50,000 inhabitants (Hobsbawm, 1968). By 1911, almost 80 per cent of the population of England and Wales lived in towns of more than 5,000 inhabitants (Wohl, 1983). The second half of the nineteenth century had seen the urban-industrial proletariat come to predominate as the main social form of the subordinate classes (Hobsbawm, 1984).

One disabling feature of the industrial city was the new separation of home and work, a socio-spatial phenomenon which was all but absent in the feudal era. This disjuncture of home and work created a powerfully disabling friction in everyday life for physically impaired people. In addition, industrial workplaces were structured and used in ways which disabled 'uncompetitive' workers, including physically impaired people. The rise of mechanised forms of production introduced productivity standards that assumed a 'normal' (that is to say, usually male and non-impaired) worker's body and disabled all others. As Ryan and Thomas note, the coming of industrialism meant the end of paid work for many disabled people who had formerly been integrated into domestic production:

> The speed of factory work, the enforced discipline, the time-keeping and production norms – all these were a highly unfavourable change from the

slower, more self-determined and flexible methods of work into which many
handicapped people had been integrated.

(1987: 101)

In 1835, Andrew Ure noted that the object of the new factory discipline was to
train formerly independent workers to 'renounce their desultory habits of work'
in order that they might 'identify themselves with the unvarying regularity of the
complex automaton' (Ure, 1967: 13). By enclosing labour power in factories,
employers were able to subject workers to a uniform set of requirements concerning punctuality, the numbers of hours and days worked, and the application
of effort (Adas, 1989). These imperatives were commonly enforced through
strict time-keeping, fines (for lateness, disobedience and slow work), the regi-
mentation of movement, and the observance of rigid performance standards
(Doray, 1988).

The regularisation of labour rhythms presupposed that labour power was supplied in a common, non-impaired form (Rabinach, 1990). The manufacture of
glass at a French glass-works in the 1860s is a typical example of this presumption
in action, with tasks in the factory in question being based upon:

> forms of co-operation requiring physical strength (as when ten workmen
> had to carry a sheet of glass weighing 300 kilograms with perfectly synchro-
> nised movements).
>
> (Doray, 1988: 13)

The law of value could not have operated effectively in factory production without this new labour regimen. A universal work discipline was necessary for the
setting, and enforcement, of average labour times. Once in place, the industrial
work process could be used to drive average labour times progressively lower.
Pollard reports one historical example where an employer – perhaps in view of
the callowness of his labour power – used incentives, rather than punishments,
to stimulate a downward trend in average labour times:

> at one silk mill, employing 300 children aged nine or less, a prize of bacon
> and three score of potatoes was given to the hardest working boy, and a doll
> to the hardest working girl, *and their output then became the norm for the
> rest.*
>
> (1963: 266) (emphasis added)

Thus, as Marx (1981) pointed out at the time,[3] industrialisation and urbanisation produced an 'incapable' social stratum, a mixed estate that could not sell its
labour power at the average rate of productivity, and which was therefore con-
signed to the usual consequences of labour market exclusion: poverty, ill-health,
brevity of life, socio-spatial marginalisation, and, for many, dependence upon
the informal sector of the economy. We might include here, as Marx did, widows, the elderly, orphans, the sick, and, interestingly, those individuals he called

'the mutilated and the victims of industry', in a clear reference to disabled people. He referred to this heterogeneous social group as the lumpenproletariat. By
the late nineteenth century in most industrialised nations, many among the 'incapable' had been incarcerated in what Foucault (1979: 199) termed 'the space
of exclusion', a new institutional system of workhouses, hospitals, asylums, and
(later) 'crippleages', operated by an extensive private charitable sector and a host
of local and central state bodies.

Dorn (1994) and Davis (1995) both show how a range of cultural and institutional forces acted to construct powerful notions of corporeal normalcy/deviancy
around the impaired/non-impaired dichotomy. These cultural material constructions
– reinforced increasingly through state practices – served to stabilise and reinforce the political-economic devalorisation of impaired labour power. Indeed,
much of the social authority gained by medicine during the nineteenth century
was achieved through its success in promulgating normalising discourses around
the body. Medicine acted in concert with the emergent discipline of statistics to
'explain' how inability to labour was in fact a 'natural' consequence of physical
deviancy. As Foucault (1979) observed, the conjunction of these political-
economic and cultural-institutional forces served to reduce the body as a political
force by disciplining its inherent, unruly heterogeneity, while also maximising
the body as an economically useful force through its enslavement to industrial
rhythms.

Apart from 'feebleness', physical deviancy was also eventually linked through a
variety of medical and pseudo-scientific discourses to a range of other dysfunc-
tional social attributes, such as vagrancy and criminality. Davis (1995) also shows
how other cultural discourses and practices – notably, popular literature and
entertainment (such as circuses) – began to reflect and fortify scientific stereotypes of corporeal normality. Of course, as Engels' (1973) gruelling 1844 survey
of Manchester's lower orders showed so vividly, the co-dependent ideals of physical normality and economic independence lay well beyond the reach of most in
the lumpenproletariat, whose 'weak bodies' (to recall Marx) were no protection
against the harsh expectations of industrial capitalism.

Marx observed that the lumpenproletariat's only alternative to dependence
upon public and private (including family) 'benevolence' was a wretched, inse-
cure form of independence, based 'on kinds of work that can only count as such
within a miserable mode of production' (1981: 366). Among the 'miserable'
jobs Marx was referring to were the many street trades – hawking goods and
services to passers-by – which made the thoroughfares of the industrial city *sites*,
rather than merely *conduits*, of economic production.

For disabled people, their economic devalorisation, or 'incapability' to
use Marx's term, took on a particular socio-spatial form. This can be understood dynamically, as a general tendency, *but not law*, for marginalisation to
specific realms of the city (and sometimes beyond). The motive forces for
this were both centrifugal and centripetal in nature, and were sourced in the
three key sites that framed the 'social space of impairment', home, workplace
and institution.

The urban social space of impairment

The paid workplace was the principal centrifugal force of marginalisation: the site where the devalorisation of disabled labour power was actually practised; a node of repulsion for disabled people. The economic centrality of that exemplary site of industrialism – the factory – gave this centrifugal force an urban, and therefore social, significance that cannot be underestimated. The key centripetal site for disabled people was the institution. The increasing ubiquity during the nineteenth century of this deliberate, and morally instructive, caricature of the factory (Foucault, 1979) meant that its tentacles reached to most corners of proletarian social space, drawing in redundant labour power for storage in institutional warehousing.

The third key element of social space for disabled people was the proletarian home. Domestic spaces were certainly important sites for the physical genesis of impairment (though, as Marx and many other Victorian commentators observed, it was the factory which produced physical impairments on an industrial scale). However, the home was an ambiguous site: many households quickly, and without sentimentality, rejected their disabled members, either for the institutions or the streets. At other times, in the context of affective domestic relations, disabled people were able to resist the centrifugal and centripetal currents of industrialism. Many Victorian working-class families harboured disabled relatives, sometimes in a tug of war with the poorhouse.

For the so-called 'incapable stratum', homeworking was one common strategy for transcending the centrifugal tendency of the factory to utter devalorisation. The move to factory production encouraged a rise in homeworking, especially in the clothing and footwear industries (Pennington and Westover, 1989). Other home industries that flourished (usually as adjuncts to factory production) involved the making-up of small items, such as matchboxes, parasols, flowers, brushes, sacks and cardboard boxes, to name but a few of the plethora of outworking activities. Many, if not most, homeworkers were subjected to 'sweating', which meant low pay and long hours.

Homeworkers were paid through low piece rates which meant that 'slow-workers' were employable. Industrialists could evade the regulatory sphere of the Factory Acts, which only covered the enclosed domains of labour, to tap a reservoir of cheap, non-militant labour power, in the form of women, children and impaired persons, for certain unskilled tasks. Homeworking thus must have been a resort of many impaired persons anxious to contribute to their household's subsistence. The Fabian socialist B.L. Hutchins observed in 1907 that sweated workers endured this form of exploitation either by 'reason of sex, age, *infirmity*, or want of organisation and support' (quoted in Pennington and Westover, 1989: 101) (emphasis added). Being, perhaps, more interstitial than marginal, homeworking must have been a relatively common coping strategy for impaired people confronted with an increasingly disabling paid work environment.

This three-way typology of workplace, institution and home helps to frame the social space of impairment in the industrial city. But what of the street? How does it fit into this social space? Was it simply a conduit that carried the centripetal and centrifugal currents of social power between these key spatial nodes? Was it an important site for disabled people?

In a concrete sense the street – and not just the slum lane – was certainly important to disabled people. Analysis of the nineteenth-century urban commentaries indicates that disabled people were a common sight on the Victorian city street, particularly in major pedestrian thoroughfares. I say 'sight' rather than simply 'inhabitant' because disabled people were distinguished from the masses of pedestrians: first by the social inscriptions of difference arising from their apparent disablement, and second, by the nature of their presence on the streets. In the various tableaux of cities constructed by journalists and literary writers, for example, the disabled beggar or trader is usually an element within the kaleidoscopic backdrop of furious, modern streetlife (Brown-May, 1995). One 'underworld journalist', Thomas Archer, journeyed through the visceral realms of the slum and the workhouse during the 1860s – many disabled people figure within his rich portrayals of urban squalor (Archer, 1985). However, disabled people were rarely foregrounded in contemporary urban descriptions, such as Archer's; almost never were they given voice.

Disabled people were not 'pedestrians'. For one, their frequently restricted ambulatory status made for a different type of participation in streetlife. Moreover, disabled people were often on the street for very immediate economic reasons, engaging in either begging or petty street trading, thus distinguishing them from strolling consumers, people in circulation, idlers or others for whom the street was not the immediate source of their existence. The street was a place of subsistence as much as it was a stage that constantly retold the story of their social difference and exclusion.

If disabled people were present on the Victorian city street, as both agents of petty commerce – street traders – and symbols of anti-commerce – beggars – we might say that this indicates both the failure and success of the oppressive structures which bore down on them. How so? By clinging to society on the streets, some disabled people resisted the 'duty to attend the asylum' (as Foucault would have it), that weighed increasingly heavily upon them as the century progressed. Alternatively, others, those remembered as 'crippled beggars', were indeed a public revelation of the crushed and lowest stratum of industrialism. For these, the street was really a 'non-place', the site of a truly wretched existence which served only as a waiting room before the inevitable moments of institutionalisation and/or death.

Abject bodies

The word 'abjection' describes both the action of casting off, of excluding a person or group, and the experience of being cast down, of degradation. Drawing upon Kristeva's (1982) work, Sibley has developed a geographic notion of abjection as both the 'unattainable desire to expel' those things which threaten the socio-spatial boundaries of normality and 'that list of things and threatening others' (1995: 18). Sibley's notion of abjection illuminates the experience of disabled people in the nineteenth-century capitalist city. Through their 'incapability' disabled people threatened the Victorian social order which was framed by economic class structures. However, both the diffuseness of oppressive power and the determination of many disabled people to resist exclusion, meant that these 'threatening others' could not be totally expelled from the public view – from city streets – and placed within the safe institutional boundaries of the 'space of exclusion'. The presence of disabled people on the Victorian city street was, as Sibley (1995) would have it, *a ritual of abjection*, a sort of uneasy (and unstable) truce between the oppressor and the oppressed. The idea of corporeal abjection here echoes Dorn's (1994: 154) notion of 'dissident bodies', which he defines as those forms 'that are particularly resistant to the articulated norms of the locality' in which they are placed.

Dorn (1994: 14) believes that 'Europe in the seventeenth century produced a public sphere stripped of the grotesque'. There is certainly an important truth in this observation – the sanitary and planning reform movements in late nineteenth-century Britain sought to impose upon public urban spaces, including streets, notions of order and cleanliness derived from the medicalised discourses of physical normality. Increasingly, public areas were no place for deviant bodies, whose presence might threaten the moral outlook and health of the normal populace. None the less, I think Dorn (1994) neglects here the ways in which disabled people – the abject, grotesque bodies of industrialism – managed to resist this 'sanitisation' of the public sphere for a long time by clinging to interstitial public spaces, such as the street and fairgrounds. Dorn himself notes how the displays of grotesque bodies in circus sideshows preserved a residual glimpse of corporeal deviancy for the wider public. In the case of the street, the reality of abjection – so powerful in the sideshow – was tempered by the fact that many disabled people achieved here a degree of socio-economic independence.

The street, then, was a place where abjection was experienced and resistance practised against the forces of abjection. It did not simply conduct the charges of centripetal and centrifugal power that sought variously to expel and attract disabled people. Rather, the street reveals these modalities in tension with the various biographies of disabled people, with their different capacities for resistance and subversion. It was, roughly speaking, a public equivalent of the home, which was also a site of resistance and abjection.

In view of the importance of the industrial city's street to disabled people, I think it helpful to amend the typology of social space first presented in Chapter 4 in order to better reflect the empirical reality of this historical setting. This produces a new conceptual prism of social space (Figure 6.1) that helps refract some understanding of disablement from the various images of Victorian street life which have been left to us. In general it seems that the street was a place of both abjection and resistance for disabled people, but maybe not much in between.

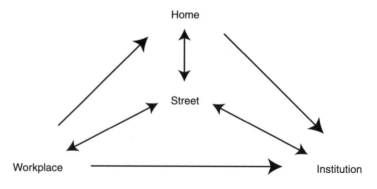

Figure 6.1 The social space of disability in the industrial city

Colonial Melbourne

'The Manchester of the Southern Hemisphere'

Colonial Melbourne can be situated squarely within the broad context of British capitalist development in the nineteenth century. The city's colonial political economy was dominated by the local bourgeoisie's unceasing efforts to emulate in almost every way the process of industrialisation followed by the imperial parent.[4] Cultural colonialism rested upon a dual subordination of indigenous peoples and the officially sanctioned cult of Anglophilia. Aboriginal cultures were quickly supressed, and in some instances annihilated, after white settlement, while, from the start, the colonial plutocracy was obsessed with mimicry of British cultural forms. While mediated by colonial circumstances, the success of this emulation in shaping Melbourne as a recognisably British, industrial urban landscape was noted approvingly by many local and foreign observers (though others decried it). In the inner municipality of Collingwood, for example, local boosters crowed that they had created a 'Manchester of the Southern Hemisphere' (Barrett, 1971). The boast was a general one, echoed by other local elites – many of them British migrants – who bragged about the creation of industrial capitalist terrains that emulated the places from which many had originally fled.[5]

Three views of social space

What was the social space of disabled people in this colonial metropolis? More specifically, how did disabled people experience the city's streetlife? My analysis of the social space of disability in colonial Melbourne showed that disabled people from the city's lower economic strata experienced socio-spatial marginalisation, meaning that many were forced to undertake marginal economic activities, such as street trading, in order to avoid forms of public and private dependency, or worse. I draw here upon three representative data sources each of which sheds light on the three distinct dimensions of the social space of impairment – workplace, institution and home (see Appendix).

The first window on this social space was the set of factory records left by Guest and Company, a large biscuit and cake manufacturing concern whose principal plant was located in the CBD for most of the Victorian period (Figure 6.2). These records – principally the employment engagement books (1889–91) – reveal both a mechanised labour process that enforced an average productivity

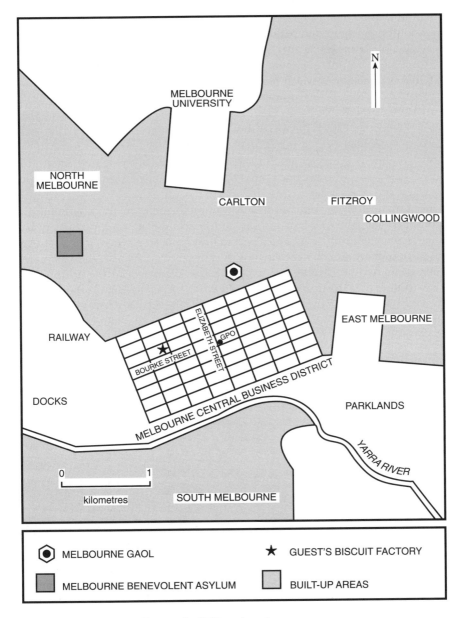

Figure 6.2 Colonial Melbourne's CBD and environs

standard and the employer's intolerance of 'slow' and impaired workers. There are many recorded dismissals of workers for being 'too slow', 'useless', 'careless', and 'unsteady'. Speed, dexterity and obedience were demanded of the workers. There is one recorded instance in which impairment is cited as a reason for dismissal in the study period. On 4 June, 1889, the foreman noted the departure of a 15-year-old boy with the following remark: 'no good, paralysed hand'. It is doubtful that the labour process at Guest's ever admitted impaired workers, at least not for any significant period. As McCalman (1984: 31) has observed more generally, Melbourne's 'factory system had no place for slow workers'.

Apart from the internal labour rhythms of the factory, there were pressing external considerations which prevented most disabled people from joining industrial labour forces. Analysis of the Guest and Company employee data for the early 1890s showed that the average daily return journey from home to workplace was six kilometres. Clearly this sort of mobility expectation was a further exclusionary factor for disabled people that added to the internal centrifugal force of the labour regime. We may assume that disabled people hardly figured among the hordes of workers who streamed through inner Melbourne's streets at the beginning and end of each day.

My examination of Melbourne's institutional landscape also confirmed the centripetal pull of the asylum and the poorhouse. By late century, the city had seventeen major institutions, and a host of other smaller places of indoor charity. A source of great pride for the city was its showpiece poorhouse, the Melbourne Benevolent Asylum (MBA), located just a few kilometres to the north of the CBD (Figure 6.3). Institutions such as the MBA helped to support the ideological construction of Victoria as an affluent and civilised land, where even paupers got to live in palatial homes (Gleeson, 1995b). But the reality was that conditions in the asylums were nothing less than barbarous. Life for those impoverished disabled people who were 'lucky' enough to gain admittance to 'the benevolent' was generally wretched and short.

Many thousands of individuals passed through the gates of the MBA after they opened in 1851: some many times, as though through revolving doors; others only

Figure 6.3 The Melbourne Benevolent Asylum, 1900

once, the moment of institutionalisation being their permanent exit from public space. The Asylum was above all a place of social and physical death. My analysis of the admissions records for the MBA between 1860 and 1880 revealed that a substantial number of its inmates at this time (597) were disabled (Table 6.1).

Table 6.1 Numbers of physically impaired males and females by stated impairment type, 1860–1880

Impairment type	No. of males	No. of females	Total persons
Disabled[a]	14	6	20
Loss of limb(s)	11	7	18
Palsy	2	3	5
Hemiplegia	1	2	3
Paraplegia	4	0	4
Paralysis	264	52	316
Part. Paralysis[b]	31	6	37
Impairment to			
Shoulder	1	0	1
Arm(s)	9	4	13
Hand(s)	7	3	10
Spine	20	16	36
Side	26	11	37
Hip(s)	17	9	26
Leg(s)	44	8	52
Knee(s)	6	8	14
Foot/Feet	4	1	5
Total impaired persons	461	136	597

Source: Melbourne Benevolent Asylum Registers of Applicants and Inmates, 1856–90
Notes:
[a] Includes those inmates described as 'disabled', 'crippled', or 'lame'.
[b] Includes those cases described only as 'partially paralysed'. In the case where paralysis of a specific part of the body was indicated, the observation has been included in the relevant impairment category.

As the city's institutional archipelago grew, so too was the heterogeneity of street life progressively narrowed, as more sub-classes of the lumpenproletariat were confined within its carceral landscape. The first, and enduring, receptacle for 'street refuse' was the gaol. There is evidence that the judiciary had a regular practice of sending street vagrants, many of them described as 'cripples', to prison as a humane gesture aimed at providing sustenance to the 'physically incapable' (Lynn, 1990).

'Helpless' persons – those among the poor with disabilities or psychiatric illnesses – were frequently imprisoned by judicial authorities for want of institutional alternatives. The usual pathway from the street to prison for the disabled poor was via an arrest for vagrancy. This was not always intended as a punitive measure; police commonly used their powers under the vagrancy statute on compassionate grounds when a needy indigent was brought to their attention. The problem was that, in the face of perennial institutional crowding, magistrates had little option but to commit the vagrant poor to the city's gaols. In 1863, the author of an anonymous letter published in a major city newspaper (very possibly a Member of Parliament) stated the problem succinctly:

> Victoria might well be proud of her public institutions, considering her youth … However much our Government has done, there are some unfortunate classes unprovided for. The maimed, the diseased, and the unfortunate widows and destitute children are insufficiently cared for. It is scarcely right that an unfortunate cripple should be treated as a vagabond, and sent to prison under the Vagrant Act, merely to provide sustenance. Yet it is the only humane way for the bench at present. It is really too bad that no comprehensive legislative measures have been made to provide for those who are physically incapable of earning their living.[6]

The British pattern of bourgeois 'assistance' to the lower classes described above was largely replicated in the colonial setting. The voluntary organisations with the most intimate knowledge of proletarian domestic space were the ladies' benevolent societies. These associations, commonly with Protestant evangelical ties, consisted of 'lady visitors' who visited the homes of needy working-class families, dispensing limited assistance and boundless 'advice'. Members were recruited from women of the lower or middle strata of the bourgeoisie, usually the wives of doctors, businessmen, and minor clergymen. Colonial Melbourne had 26 such ladies' benevolent societies by the 1880s (Kennedy, 1985).

Among these, the largest and most influential was the Melbourne Ladies' Benevolent Society (hereafter, MLBS or 'the Society'). Throughout the second half of the nineteenth-century, the Society was the principal source of outdoor – that is, 'home delivered' – charity in the city. An important third window on the social space of disability is provided in the voluminous case records left by the MLBS.

From 1855, the Society's field of operation settled upon the CBD and four adjoining suburbs; the whole divided into forty smaller districts, each with its own lady visitor (Figure 6.4). By the 1890s, the Society's operating area was home to about 150,000 persons. These areas contained extensive slum tracts and a considerable lumpenproletariat of widows, families deserted by a breadwinner, the aged, the sick and disabled people.

The Society's lady visitors have left us extremely rich accounts of the domestic and public life of the industrial working class. The records covering 1849-1900 reveal that the Society aided many families with disabled relatives: among these, I was able to identify 1,004 disabled individuals (doubtless an undercount) (Table 6.2). Interestingly, the impairment types closely paralleled those recorded at the MBA. Indeed, a few individuals figured in both sets of data, indicating the centripetal pull between the institution and home. Meagre though it was, outdoor charity such as this helped many families and disabled individuals to resist the centripetal pull of the institution, if only for brief periods.

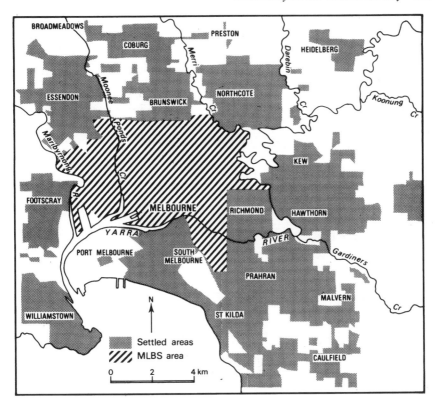

Figure 6.4 The sphere of operation of the Melbourne Ladies' Benevolent Society, 1855–1900

Clinging to the home, and sometimes family, none the less often meant exclusion from public space for disabled people. The Society records confirm that many disabled people engaged in sweated homework to sustain themselves: such people were rarely seen in the street. One of the many recorded examples was reported thus:

> elderly widow, with very bad leg, found her working at shirts, with her leg up on bed.[7]

This picture is confirmed in other surviving fragments which record the visceral world of sweating. In his 1891 report to Parliament, for example, the Chief Inspector of Factories relayed the following pathetic case of two outworkers:

> These girls live with parents, and pay them for keep ... One is a cripple and is laughed at by factory hands.[8]

The Chief Inspector remarked that the crippled sister had rarely left the home after being humiliated by factory workers.

Table 6.2 Numbers of physically impaired children and adults by stated impairment type, 1850–1900

Impairment type	No. of children	No. of adults	Total persons
Crippling condition	60	107	167
Disabling condition	3	33	36
Loss of limb(s)[a]	3	76	79
No use of limb(s)[a]	1	36	37
Weakness of limb(s)[a]	–	4	4
Deformity	8	5	13
Paralysis	10	179	189
Lameness	8	58	66
Disease of spine, etc.[b]	31	21	52
Club foot	1	1	2
St Vitus' Dance	1	–	1
Permanently invalided	–	7	7
Long term injury[c]	–	18	18
Total impaired persons	126	545	671
Possible impairment[d]	3	330	333
Total impaired and possibly impaired persons	129	875	1,004

Source: MLBS Minutes, 1850-1900
[a] Includes persons having lost part of limb(s).
[b] Includes persons with diseases of leg(s) and hip(s).
[c] Includes persons disabled for at least six months by injury.
[d] Includes persons with 'bad' limb(s) or part thereof.

Alternatively, some disabled people used the home as a base for street trading and begging as supplements to charity and the aggregate household income. The Society smiled on any attempt to please the great Victorian gods of Thrift and Independence, and it sometimes helped disabled people to establish themselves in various street trades, as the following case notes reveal.

In 1873, the Society helped a recently disabled man to establish himself as a produce hawker:

> respectable couple, husband has had a broken leg, not properly set, and is anxious to get a stand in the market.[9]

Seventeen years later, the practice was still common, as one lady visitor's report on the plight of one woman makes clear:

> husband wooden leg, and wants pounds to start with vegetables ... four children.[10]

For women in such a position, the procurement of a basket to facilitate the selling of fruit or flowers was sometimes a means to some form of economic independence:

[name] is desirous of obtaining a basket to sell fruit, one arm being disabled preventing her taking a situation.[11]

Other impaired people attempted the life of a street musician. The Society sometimes helped with the purchase of an instrument in such cases. In 1875, for example, it assisted a

widow, with a grandson, who is a cripple, and a musician – is anxious for aid towards the purchase of a flute.[12]

Some impaired people seem to have survived by combining street trading and charity with a measure of old-fashioned, venial roguishness. In March 1891, a visitor reported assisting a Fitzroy woman, whose

husband is a cripple, but has a coffee stall. They are dirty thriftless people – brawling and noisy.[13]

One can almost hear the clicking of tongues which greeted this disturbing news and it was duly recorded that aid was to be discontinued in this case. But, in practice, the ladies were rarely as stern as their recorded pronouncements, and the visitor clearly relented in this case, because two months later she was forced to make the following distressing report concerning the case of the obstreperous coffee vendors:

on visiting found everything cleared out of the house. The coffee stall was in the yard. Mrs [name] learned from Sergeant of Police that they had nothing to pay for the stall, and that a warrant was out for [name]'s husband.[14]

In my analysis of the data, I was able to conclude with certainty that thirty-six of the disabled people assisted by the Society were in some form of paid work (Table 6.3). Many of these trades were clearly either some form of street hawking or outworking.

The view from the street

Sadly, apart from these fragments, there are few surviving records of what life was like for Melbourne's disabled street hawkers. However, the commentaries of one English observer of streetlife have recorded the voices of other disabled traders, and these speak powerfully of the abjection which must have overshadowed the lives of the industrial lumpenproletariat. Henry Mayhew (1968a, 1968b), renowned slum journalist of mid-century London, was one of the few observers of the Victorian city to foreground disabled street people (all of them males), if only for a few instants. Mayhew regarded disabled people as 'one of the classes *driven* to the streets by utter inability to labour' (1968a: 329) (original emphasis). His

Table 6.3 Occupations of working impaired persons

Occupation	No.
Streethawker[a]	5
Needle woman	4
Carter[a]	2
Musician[a]	2
Organ grinder[a]	2
Writer	2
Coffee stall holder[a]	2
Parasol mender	1
Office worker	1
Washer woman	1
Flower maker[b]	1
Flower seller[a]	1
Boot finisher[b]	1
Rag picker[a]	1
Toymaker	1
Mill worker	1
Presser	1
Messenger	1
Shirtmaker[b]	1
Tinsmith	1
Knitter[b]	1
Newspaper seller[a]	1
Caretaker	1
Match seller[a]	1
Total	36

Source: MLBS Minutes, 1850–1900
Notes: [a] 'Street occupations'
[b] Possible outworkers

surveys record fragments of these unique street lives in the form of sketches, snatches of conversation, and his own commentary.

On one of his meanderings through London's slum streets and rookeries, Mayhew asked a crippled bird seller why he was working. The reply:

> Father didn't know what better to put me to ... I liked the birds and do still. I used to think at first that they was like me, they was prisoners and I was a cripple.

> (1968b: 68)

He then tells Mayhew that when his father died he succumbed to the centripetal pull of the workhouse:

> O, I hated it ... I'd rather be lamer than I am, and be oftener called Silly Billy – and that sometimes makes me dreadful wild – than be in the workhouse.

> (ibid.)

Figure 6.5 The street-seller of nutmeg-graters

His view of the future was bleak:

> I feel that I shall be a poor starving cripple, till I end, perhaps, in the work-house.

> (ibid.)

Mayhew next encounters a seller of nutmeg-graters whose plaintive appear-ance bears powerful witness to the daily ritual of abjection in which many disabled people were compelled to take part (Figure 6.5). The seller wears a prominent sign around his neck declaring 'I WAS BORN A CRIPPLE' in recognition of the obdurate suspicion of middle-class Victorians that all disabled street-traders and beggars were really well-disguised, 'healthy' vagrants manipulating the sympathies of gullible passers-by. The nutmeg-grater seller tells Mayhew that his relatives despise him for his disabilities and had abandoned him. Mayhew congratulates him for his initiative in making a 'living', and moves on.

Later, Mayhew interviews a one-legged crossing sweeper (Figure 6.6) who has little to say other than:

Figure 6.6 The one-legged crossing sweeper

A man had better be killed out of the way than be disabled.

(1968b: 488)

These scenes bear witness variously to resistance and abjection, and were doubtless reproduced daily in colonial Melbourne. Many of Melbourne's disabled street traders would never have found their way into the Ladies' Benevolent Society's records, as most were probably either homeless or nomads within the liminal accommodation sector of lodging houses, refuges, and the like.

The CBD was the scene of a vibrant bourgeois street life, famous for its shopping arcades, galleries, theatres and palatial hotels, with the city being variously named the 'Paris' or 'Chicago' of the South (Davison, 1978). The city was one of the first in the Empire to receive electric street lighting, and the nocturnal parades of intermingling bourgeois and proletarian streams were a source of both fascination and concern for Victorian commentators. The flow of streetlife would constantly negotiate the many beggars and street-traders who populated the streets of the city and adjoining areas (Kennedy, 1982). From the 1860s, a succession of modes of public transport, such as horse (later, cable) trams, evolved to carry the growing throngs, thereby increasing the velocity of street traffic by separating the strollers from the travellers. These were hardly accessible forms of transport for many disabled people.

By the 1880s, Bourke Street, an important commercial thoroughfare in the city centre, was the centre of the interstitial street economy, and was daily the host of hawkers selling everything from fruit to matches. (By night, the offerings extended to (mostly, female) bodies.) These petty merchants competed with a brigade of street musicians and entertainers for prominent positions (Kennedy, 1982); usually points of maximum friction and/or visibility in the flow of street traffic such as street corners (Brown-May, 1995). Swain (1985) relates the story of Ada, a partially blind single mother, who survived in the early 1890s by singing and selling matches on city streets until finally arrested (and separated from her child). Swain notes that, 'Ada was not atypical, for many similar girls were also physically or mentally handicapped and quite alone in the city' (1985: 99).

John Freeman, in his *Lights and Shadows of Melbourne Life* (1888), describes women beggars displaying their crippled children in order to elicit sympathy and alms. (Some are even said to have 'borrowed' impaired children for the purpose.) Freeman's prose portrait of street begging and trading also contains several references to impaired hawkers and musicians. One 1887 account records a crowd gathering on a busy street corner to watch a party of showmen: 'the chief attraction is a so called fortune teller called "Gypsy Eliza" and a deformed man' (cited in Brown-May, 1995: 28). Note the anonymity of the disabled showman.

Many among the bourgeoisie were clearly alarmed at this unregulated intercourse of classes and moral types in public streets. The Charity Organisation Society (COS), Melbourne's self-appointed guardian of proletarian virtue, was particularly concerned at the moral threat posed by this daily street carnival. The 1890 report of the COS fairly recoils at 'the spectacle of old and young, tainted and untainted, commingling and competing in the streets'.

Strange that these ruthless champions of laissez-faire should find this quintessentially capitalist assemblage so disturbing. On a more bizarre level, Andrew Halliday, Mayhew's collaborator, disguised his Victorian squeamishness with voodoo science: 'Instances are on record of nervous females having been seriously frightened, and even injured, by seeing men without legs or arms crawling at their feet' (Mayhew, 1968c: 433). Indeed,

> A case is within my own knowledge, where the sight of a man without arms or legs had such an effect upon a lady in the family way that her child was born in all respects the very counterpart of the object that alarmed her. It had neither legs nor arms.
>
> (ibid.)

In the same set of remarks, Halliday exhorts the police to prevent the attempts of 'some of the more hideous of these beggars to *infest the street*' (ibid.) (emphasis added).

Halliday's metaphorical turn suggests that crippled beggars were vermin which directly threatened public health, and points to the extreme sense of abjection that the publicly displayed disabled body could conjure in the minds of the Victorian ruling classes. Halliday, of course, was not alone in his anxieties: purification of the street became an ideal for many among the late Victorian bourgeoisie. Anxieties about disease converged with fears of social difference in a new campaign for 'street hygiene', the demand that disgusting and contaminating 'objects' be removed to remote institutional spaces in order that the public's health – moral and physical – might be properly safeguarded.

As the century wore on, Melbourne's civic authorities duly responded to these sorts of anxieties and other imperatives by tightening general controls on street trading and vagrancy, thereby heightening the centripetal pull of the institution. Brown-May notes that in Melbourne, 'Street life came to be viewed with suspicion, as deviant and pathological, demanding regulation and control' (1995: 30).

Although official intolerance of the disabled street beggar hardened as the century progressed, the reverse was probably true of the disabled street trader. Eventually, officialdom came to believe that orderly street trading was a respectable and humane solution to the problem posed by immobile and 'incapable' labour-power. Interestingly, it was the COS, Melbourne's moral guardian of charity, which encouraged the acceptance of this 'humane' view:

> In the fullness of time the COS would convert governments, local councils and the police to its viewpoint that street begging should be banned, street vending licensed, and the 'privilege' of street stalls in some locations 'reserved almost exclusively for those under some physical disability'.
>
> (Kennedy, 1985: 209)

During the 1890s, the COS 'established a crippled person in Melbourne's first newspaper kiosk for the disabled' (Kennedy, 1985: 199). This confinement of disabled street traders in well-concealed kiosks was a prelude to the sheltered workshops that were finally to remove disabled people from the public sphere in the next century.

Conclusion

The transition from feudalism to industrial capitalism was achieved only after long, conflictual struggles that eventually re-shaped the social space of Europe's lower orders. By the early nineteenth century, feudal rural society had largely given way in England to an increasingly urbanised, proletarian lifeworld, where the rule of commodity relations overshadowed, and eventually supplanted, many traditional cultural forms of association. The new industrial, political-economic order devalorised impaired labour power, a shift in social embodiment that was reinforced by emergent cultural-institutional constructions of normality and social dependency.

The social space of *disability* that arose from these broader changes is best conceived as a dynamic, restless landscape marked by centripetal (inclusionary) and centrifugal (exclusionary) pressures on impaired bodies. Specifically, this space was characterised both by the continual displacement of disabled people from social mainstreams, and by their efforts to resist this tendency to marginalisation. Displacement forces were sourced in three key socio-spatial nodes: centrifugal pressures emerging from the home and workplace (especially the factory); and the centripetal pull of the institution. Resistance to the vectors set by these flows of power was always present: the most successful counteraction probably occurred in the home where a combination of factors – notably, the presence of affective ties and the possibility of domestic piece work – provided the resources for resistance.

The street provided a further, interstitial place where resistance was practised by disabled people. Their recorded presence in public thoroughfares exposes both the frequent practice of insubordination against exclusionary powers, and also the inability of oppressive structures, in this case industrial capitalism, to expel entirely all 'threatening others'. Of course, the street was above all a place of abjection for disabled people, where oppression was experienced in full public display, a fact that disturbed those bourgeois reformers who thought that the interests of the 'moral order' would be better served if all devalorised bodies were safely enclosed in the proper institutional spaces reserved for them.

Disabled people took to the streets of the industrial city in a struggle for personal autonomy and social inclusion. Although quite unlike the disability social movements of recent times, these frequent, if isolated, struggles none the less constituted a form of insurgency against an exclusionary social order. Mayhew's crippled seller of nutmeg-graters evoked the determination of many disabled people to resist the dreaded institution at any cost, declaring that he would 'rather die in the streets than be a [workhouse] pauper' (1968a: 332), and it is indeed probable that he finally succumbed one day in the place of his trade. However, I think that we must not end here with a cheer for resistance, as these shadows and fragments of autonomy should not be confused with real liberation and social inclusion. For many disabled people the street was a place of both struggle and abjection, and we cannot, from this distance, remember these painful biographies without sadness. For the contemporary observer, this moment of reflection honours certain 'uneventful' lives that passed quietly, though never passively, within the industrial city.

Part III

Contemporary geographies of disability

7   Disability and the capitalist city

Introduction

In this final part of the book I address the question of disability in contemporary capitalist societies. As I explained in the introductory chapter, my interest here is in the urban context of disability, although all of the social geographic problems I will explore in this and the following two chapters also exist outside major cities. In the future, I expect that non-urban contexts will be the subject of further geographies of disability. By addressing the broad context of urban disablement, I hope to provide analyses that can be extended to non-urban settings. Of course, such analytical extension will require some reformulation of the concepts used here and their empirical reference to the distinct conditions facing disabled people in rural areas and small-scale settlements.

The aim of this chapter is to suggest an urban geography of disablement; in other words, a potential framework for understanding the oppressive experiences of disabled people in contemporary and near-contemporary Western cities. The chapter will review how disability, as a specific socio-spatial experience, is a critical feature of the capitalist city. This is to argue, *inter alia*, that disablement – the oppressive experiences of physically impaired people – is deeply inscribed in the discursive, institutional and material dimensions of capitalist cities.

The first two sections explain the phenomenon of 'disability oppression'. Here I explore the political-economic, cultural and spatial dimensions of disability oppression. After this, I briefly review how disabled people, and their allies, have opposed in a range of Western countries these tendencies through their own urban social movements that have focused on key policy issues, such as civil rights, and environmental accessibility. In contrast to the disabling form taken by contemporary Western cities, a different vision of produced space is forwarded in the final section, with the discussion offering in broad outline the requirements for 'enabling environments' – non-oppressive and inclusive social spaces. This political-ethical principle is particularly aimed at countering disabling practices and ideologies within the spheres of state policy, and will thus provide the normative basis for the analyses in the following two chapters.

Disability as social oppression

Disability oppression

From the voluminous policy and theoretical literatures produced by the disability movements of Western countries in recent years, it is clear that disabled people are confronted by a common set of social disadvantages.[1] In contemporary social science literatures, several terms – notably, 'disablism' and 'ableism' – have been coined to describe these disadvantages, and the discriminatory structures and practices that produce them (cf. Chouinard, 1997; Imrie, 1996a). Of course, physically impaired people are more than simply the sum of their disabilities, and their individual experiences of disablism/ableism are shaped in specific ways by differences in social identity and group affiliation.

In these essays on contemporary geographies of disability, I will use the term *disability oppression*, in preference to these other ways of describing the repressive socialisation of impaired bodies in Western societies. Terms such as ableism and disablism have the advantage of specifying disabled people's oppressive experiences with respect to other, well-known social cleavages. However, these very distinct descriptive terms may also, in some discursive contexts, overemphasise the singularity of disabled people's identities and, in so doing, underestimate the connections between disability and the broad currents of social oppression that arise from culture and the political economy.

By connecting the idea of disablement and social oppression, 'disability oppression' conveys the dual quality of this form of disadavantage, as a singular form of injustice that is deeply imbricated with other structures of subordination. Moreover, many disability theorists, including the materialists whose work was examined in Chapters 2 and 3, have explained disablement as a form of social oppression that is intimately connected to other forms of subjugation (e.g., Abberley, 1987; French, 1993a, 1993b; Morris, 1991; Oliver, 1990; Swain *et al.*, 1993). Oliver (1996) reflects the view emerging among many disability commentators that disablement is one form of a broader, oppressive socialisation of embodiment that includes other 'repressed bodies' identified on sex, gender and race lines.

Five faces of disability oppression

Young's (1990) notion of oppression as a multi-dimensional phenomenon is useful in helping to clarify the singular (particularly, *economic*) nature of disability as a form of social discrimination.[2] Harvey (1993) has very usefully paraphrased her five-way typology of oppression as follows:

1   Exploitation: the transference of the fruits of labour from one group to another, as, for example, in the case of workers giving up surplus value to capitalists, or in the case of women, in the domestic sphere transferring the product of their labour to men;

2   Marginalisation: the exclusion of people from participation in social life so that they are potentially subjected to severe material deprivation and even extermination;

3   Powerlessness: the lack of that authority, status and sense of self which would permit a person to be listened to with respect;

4   Cultural imperialism: stereotyping in behaviours, such that a dominant culture imposes on the oppressed group its experience and interpretation of social life; and

5   Violence: the fear and actuality of random, unprovoked attacks which have no motive except to damage, humiliate or destroy the person.

In a wide-ranging, but constructive, critique of Young's schema, Fraser offers an additional, indeed complementary, explanation of oppression in the form of a conceptual spectrum which posits two distinct poles of injustice for social collectivities: on the one hand, distributive injustices sourced in the political economy and therefore requiring redistributive remedies; and, at the other pole, the injustice of cultural misrecognition which is 'ultimately traceable to the cultural-valuational structure' of advanced capitalist societies, and therefore must be remedied through cultural revaluations (1997a: 18).[3] Fraser believes that the contemporary proletariat provides the clearest example of a collectivity subject to distributive injustice, while homosexual groups – or indeed, any 'despised sexuality' – best represent the case of cultural misrecognition. Of course, Fraser doubts that there exist any 'pure collectivities' of these sort, but the spectrum plays a useful heuristic role by identifying the types of oppression that most affect different groups. Indeed, her view is that both major forms of oppression – redistributive-cultural – intertwine in concrete social settings, though the strength of these interdependencies varies with the social collectivity in question.

A more complex political-ethical issue, according to Fraser, is the case of the 'bivalent' collectivity, a group located in the middle of her conceptual spectrum and thus experiencing 'injustices that are traceable to both political economy and culture simultaneously' (1997a: 19). I argue that disability is, like gender and race, a bivalent collectivity possessing all five of Young's faces of oppression. I would, however, choose to place disability closer to the redistributional end of Fraser's spectrum, for its particular embodiment of forms one and two, exploitation and marginalisation. Thus, material exploitation and physical marginalisation of disabled people are both partly achieved and reproduced through discriminatory mainstream cultures and political structures. In Western societies, disabled people are not subjected to open and systematic violence, at least not in the manner experienced frequently by racial or ethnic minorities. However, the inhumanity that still characterises the services delivered in many institutional settings, and the brutality of certain medical treatments for impairment, are surely both instances of violent behaviour towards disabled people that is still practised in developed societies. Moreover, disabled people are frequently the victims of sexual abuse and physical violence in institutional and community based residential facilities.

In the following two sub-sections, I compose a portrait of the general context of oppression confronting disabled people in contemporary Western societies. After this, I consider how this bivalent oppression takes on a particular form in cities.

Economic oppression

The disabling division of labour

The international social sciences literature has established that disabled people frequently lack many of the basic material and cultural resources that are needed to sustain both a healthy existence and social participation (Alcock, 1993). The oppression confronting physically impaired people can be distinguished from other forms of socio-economic disadvantage, having unique characteristics in terms of labour market participation, physical access, social service use, income levels, and political participation.

Physical impairment imposes a distinct set of economic costs for individuals (Berkowitz and Hill, 1989), many of which – e.g., transport expenses – are aggravated by discriminatory forms of social organisation. These 'costs of disability' can include payments relating to the use of specific medical, social and transport services needed by impaired persons, as well as expenditure on personal appliances and accessories. Such costs – for which there is often no recompense from governments or other bodies – can radically reduce the net income of those disabled people who are fortunate enough to obtain paid employment. Indeed, this reduction can discourage many disabled people from even trying to enter formal employment markets. Moreover, disabled people may suffer two perverse consequences from paid employment: first, they may receive low, or even negative, net wages after costs are deducted; second, employment may reduce the eligibility of disabled people for supporting pensions and other welfare or health related payments.

In the event, many employers are unwilling to engage physically impaired people, fearing that such workers would be unproductive and/or have vocational needs that would disrupt workplace rhythms. While such individual fears are frequently baseless,[4] they also neglect the wider issue, raised by many disability commentators (e.g., Oliver, 1991; Pati and Stubblefield, 1990) that competitive commodity labour markets systematically undervalue the work potential of disabled people. Liachowitz (1988) and Oliver (1991) both see a historical link between this devalorisation and the growth of urbanisation in nineteenth-century Europe. As I argued in the previous part of the book, the motive force for this urbanisation was the rise of competitive capitalism, a mode of production which fashioned workplaces, and entire cities, around industrial labour markets that excluded 'slow' or 'incapable' workers. The economies of contemporary capitalist cities thus reveal a legacy of discriminatory industrial labour markets by continuing to valorise non-disabled labour power over all other forms.

Rates for both labour market participation and paid employment are very low among disabled people in Western countries. In Britain, for example, it was estimated in the late 1980s that only 31 per cent of the country's two million disabled people of working age were in paid employment (Oliver, 1991; see also Barnes, 1992a; Jenkins, 1991). Remarkably, a 1994 national survey in the United States found the same employment rate (31 per cent) among disabled people aged 16 to 64 years (*New York Times*, 23 October 1994: 18). According to Pati and Stubblefield (1990), there are around ten million unemployed disabled citizens of the United States who are capable of work (see also Berkowitz and Hill (1989)). Figures cited by Barnes (1992a) show that during the 1980s disabled people in Britain were three times more likely to be unemployed than non-disabled people.

The data for Australia are equally instructive. In 1993, the estimated labour force participation rate of 'handicapped' people aged 15–64 years was 46 per cent, compared with a figure of 74 per cent for the total population (Australian Bureau of Statistics, 1993). Of those handicapped people in the labour force, 21 per cent were unemployed, compared with 11 per cent of the national workforce.[5] This means that many disabled people remain dependent upon families and affective networks for care and support. The task of providing care for a dependent disabled partner, parent or child means that many carers must leave the labour force – the 1993 participation rate for principal carers was only 60 per cent. Moreover, the burden of care fell more heavily in this respect on women – the labour force participation rate of female principal carers was only 75 per cent that of male carers (ABS, 1995). Of those carers who remain in paid employment, many are forced to reduce their hours of work and, in some cases, their workforce status (Brown, 1996).

Theorists of race argue that there exists in capitalist societies a 'racial division of labour' which overdetermines the process of class exploitation (Fraser, 1997a). I argue that the capitalist social relations of production have also been characterised by a *disabling division of labour*. Recalling Foucault's comments in the previous chapter, I would describe this division as a historically specific mechanism that has sought to maximise 'the body' as an economic force by disciplining its inherent, unruly heterogeneity, while also ensuring its enslavement to industrial rhythms. This has involved the historical devalorisation of the labour power of disabled people, a situation that continues to the present day.

Thus, for the relative few who do find employment, exploitation frequently awaits in sheltered workplaces or low-paid jobs in open employment settings (Alcock,1993). In Australia, for example, it was estimated in 1990 that 53 per cent of sheltered employees were earning less than A\$20 per week (Ronalds, 1990). Further British figures show that disabled men in full-time work currently earn almost a quarter less per week than their non-disabled equivalents. Even worse, disabled women workers earned almost a third less per week than did disabled men workers, suggesting – not surprisingly – that gender plays an important role in determining the relative oppressiveness of disablement (Barnes, 1992a). There is a striking parallel in Canadian data which show that in 1991 'women aged 15–34 with disabilities earned only 68.7% of the income earned by disabled men in the same age range' (Chouinard, 1997: 381).

Poverty and disability

Largely as a consequence of labour market exclusion and work exploitation, disabled people tend to be poorer than many other socially disadvantaged groups (Oliver, 1991). Alcock (1993: 175), citing Groves, observes that 'Poverty is disability's close companion'. In support of this, he reports that three-quarters of disabled adults in Britain rely on state benefits as their main source of income. Furthermore, Alcock argues that the various 'costs of disability' can greatly depress the standard of living of this class of welfare recipients. A study by Berthoud *et al.* (1993) found high rates of poverty among disabled people in Britain, especially among recipients of various welfare benefits. Using a standard of living measure that, *inter alia*, accounted for the costs of disability, these authors found that some 45 per cent of disabled people surveyed fell below the poverty line. Moreover, Berthoud *et al.* (1993) concluded that, although disabled people in Britain relied heavily upon welfare services, these benefits were insufficient to allow most recipients to meet their costs of living. Chouinard (1997) reports that in one Canadian province, Ontario, some 80 per cent of disabled people were judged in one survey to be poor.

Glendinning (1991) argued that a decade of New Right social policy since the election of the Thatcher government had greatly reduced the quality and extent of health, employment and welfare services for disabled people. By contrast, the comparative analyses of Burkhauser (1989) and Hurst (1995) suggest that disabled people in Nordic and some Western European countries (notably, the Netherlands) fared much better than their British counterparts during this period due to the relatively generous social security provisions and employment policies in those states. Furthermore, Lunt and Thornton's (1994) comparison of national employment policies for disabled people during the 1980s applauds the initiatives of Australian and United States governments, but is generally critical of British programmes and legislation. Clearly the extent of policy and service deprivation for disabled people varies significantly between countries and over time.

Powerlessness and cultural imperialism

The social devaluation of impairment

Cultural devaluation is a major dimension of the impoverishment experienced by disabled people. To adapt Fraser's (1997a) discussion of heterosexism (another bivalent form of oppression), disabled people suffer from authoritative constructions of cultural and political norms that privilege non-impaired forms of embodiment. Since the rise of capitalism, ableist cultural norms have become institutionalised in the state, civil society and economy.

In the previous chapter, it was shown how the early capitalist state developed and applied disabling constructions of dependence/independence in new welfare and social policy realms. To these must be added public education, a key institution to which disabled people have long been denied access in most Western countries (Harris *et al.*, 1995). This form of cultural exclusion is reflected in the generally low levels of educational achievement among disabled people in Western countries – in 1991, for example, Canadian disabled adults were only half as likely as those without disabilities to possess a university degree (Chouinard, 1997). In Australia, a recent government inquiry found evidence of 'harrowing' discrimination against disabled students in mainstream schools (*Canberra Times*, 4 April 1997: 1).

In the main, official constructions of ability and economic independence have sought, with varying degrees of success, to render disabled people powerless and entrench their material dependence upon the state. With the rise of the Welfare State in the second half of this century, the disability–dependency equation was set within a broader regime of 'humane care' for service dependent people (Pinch, 1997). In recent decades, the welfarist model of care has been challenged by disabled people's movements in most Western countries. These criticisms will be further discussed later in this chapter, and in the next two chapters.

The disabling imaginary

Shakespeare (1994) has argued that proponents of the social model of disability have neglected the issue of cultural representation. His review of imagery and impairment in Western societies demonstrates how historical and contemporary cultural forms – especially literature, cinema and the popular media – have perpetuated disabling representations of the 'normal' body (see also Barnes, 1992b; Dorn, 1994; Gartner and Joe, 1987; Hevey, 1997; Holden, 1991; Ingstad and Whyte, 1995; Morrison and Finkelstein, 1993).[6] By promoting oppressive stereotypes of impairment – such as the freakish, helpless or heroic cripple – these cultural representations undermine the self-esteem of disabled people and strengthen social prejudices towards corporeal 'abnormality' (Thomson, 1997). These reductive stereotypes that pervade popular consciousness also deny to disabled people the complex and enriching reality of their multiple social identities as sexed, gendered and racialised bodies. Of course, disabling representations also reinforce the economic devalorisation of impaired labour power, and the distributional injustices that flow from this. In some cities, disabling stereotypes were codified in laws that restricted the access of physically impaired people to public space. Until recently, many United States cities had 'ugly laws' that banished disabled people from public spaces (Gilderbloom and Rosentraub, 1990). The Chicago ordinance read:

> No person who is diseased, maimed, mutilated or in any way deformed so as to be an unsightly or disgusting object or improper person to be allowed in or on the public ways or other public places in this city shall therein or thereupon expose himself to the public view.
> (cited in Gilderbloom and Rosentraub, 1990: 281n)

While such proscriptions may seem bizarre and offensive to the contemporary sensibility, the disabling imaginary none the less continues to inform discriminatory representations and practices that have taken on new forms. It is now through popular cultural forms – rather than through physical public spaces – that the abjection of the disabled body is communicated to society.[7] As Shakespeare puts it, the many stereotypical images of impairment evident in contemporary culture seem to betray a societal fear of physical frailty and bodily heterogeneity:

> People project their fear of death, their unease at their physicality and mortality onto disabled people, who represent all these difficult aspects of human existence ... Disabled people are scapegoats. It is not just that disabled people are different, expensive, inconvenient, or odd: it is that they represent a threat – either ... to order, or, to the self-conception of western human beings – who, since the Enlightenment, have viewed themselves as perfectible, as all-knowing, as god-like.
>
> (1994: 298)

This cultural antipathy of 'unruly difference' is sourced, at least partly, in the historical development of the capitalist political economy and bourgeois social institutions. As Foucault explained, these social forces have colluded to discipline the inherent heterogeneity of the human form in order to maximise the body's political docility and economic utility.

Disabling representations of impairment are examples of what Ruddick (1997), following Lefebvre (1991), has termed the 'social imaginary'. A socially constructed form of popular (un)consciousness is imaginary in the sense that it is not a straightforward reflection of the object to which it refers, which may be an excluded social group, such as the homeless youth in Ruddick's study. Rather, the 'imaginary' is produced by the discourse that surrounds the object in question. Such social imaginaries are never permanently fixed, and are continually renegotiated (Iveson, 1997). Disabled people themselves have in recent years contested the abject representations of impairment that have long pervaded mainstream culture. One example of this resistance has been the reappropriation and revalorisation by disabled people of abject terms for impairment, such as cripple, or 'crip'. In one recent example of this reappropriation, a documentary film made in the United States celebrated 'crip culture' by exploring the struggle of disabled people to gain access to mainstream cultural institutions. In Britain, the 'militant crips' of the Direct Action Network attend protests outside Whitehall in their 'Piss on Pity' T-shirts, 'sticking two fingers up at the traditional charity-campaign image of disabled people as quietly respectable, submissive types' (Daniel, 1998: 22).[8]

The notions of exclusion and marginalisation which are at the core of disablement are inherently geographic, suggesting socio-spatial boundaries and margins. In seeking to explain disability, I think we must approach these dynamics of oppression as *socio-spatial* phenomena. In the next section I discuss the urban context of disability oppression.

The disabling city

Urban oppression

Disability oppression takes a distinctive form in cities. Certain general urban characteristics – notably city design, urban employment patterns and the distribution of land uses – entrench social discrimination against disabled people. Disabled people, their advocates, and occasionally governments, have identified two main urban dimensions of disability oppression: physical inaccessibility and socio-spatial exclusion in institutionalised forms of social care.[9] While these aspects of oppression take specific socio-spatial forms in different cities, they none the less have a common genesis in the economic and cultural devalorisation of disabled people in capitalist societies. As new geographies of rural and regional experiences of disability emerge, it will be possible to better elaborate how these broad structures of oppression condition the production of space in distinct contexts. As stated in Chapter 1, my interest here is in how disability oppression is manifested in large urban areas.

Physical inaccessibility

A powerfully disabling feature of capitalist cities is their inaccessible design (Imrie, 1996a). This means that the physical layout of cities – including both macro land use patterns and the internal design of buildings – discriminates against disabled people by not taking account of their mobility requirements. Practically speaking, this discrimination takes the form of:

- physical barriers to movement for disabled people, including broken surfaces on thoroughfares (streets, guttering, paving) which reduce or annul the effectiveness of mobility aids (e.g., wheelchairs, walking frames),
- building architecture which excludes the entry of anyone unable to use stairs and hand-opened doors,
- public and private transport modes which assume that drivers and passengers are non-impaired, and
- public information (e.g., signage) presented in forms that assume a common level of visual and aural ability.

The above list is not exhaustive but does point to some of the more common discriminatory aspects of the built environments of contemporary Western cities.

Even allowing for the distinctive morphologies, economies, cultures, and planning policies of Western cities, the international breadth of concern raised by disabled people concerning inaccessibility demonstrates that this is a pervasive feature of urban life. As Hahn observes: 'In terms of ease or comfort, most cities have been designed not merely for the nondisabled but for a physical ideal that few human beings can ever hope to approximate' (1986: 273).

For disabled people, these pervasive mobility handicaps are more than simply the quotidian urban frictions which irritate non-disabled people (e.g., public transport delays, road blockages, freak weather, periodic crowding). Rather, discriminatory design is a critical manifestation, and cause, of *social oppression* because it reduces the ability of disabled people to participate fully in urban life. More particularly, mobility constraints in the contemporary capitalist city are serious impediments to one's chances of gaining meaningful employment, and hence are linked to heightened poverty risk. In addition, an inaccessible built environment reduces disabled people's capacity to both engage in political activities and establish and maintain affective ties. It is not surprising therefore that Hahn (1986: 274) sees inaccessibility as a threat to 'principles of democratic freedom and equality for citizens with disabilities'.

Both Liachowitz (1988) and Alcock (1993) argue that contemporary capitalist cities both reflect and entrench disablement through their physical inaccessibility and discriminatory labour markets. Alcock (1993) draws particular attention to the link between inaccessibility and poverty, arguing that there are many 'additional costs of coping with a disability in the able-bodied world' (Alcock, 1993:188). Inaccessibility also often means that disabled people are unable to engage in mainstream consumption activities, thereby reducing their capacity to purchase goods and services at optimal prices. These goods and services include major urban consumption items, such as housing, education, transport and finance (Oliver, 1991).

Most Western governments now have in place forms of planning and building regulations which aim to prevent or at least reduce the production of inaccessible built environments and transport systems. However, as Imrie (1996a) has shown for Britain, these regulations are often poorly enforced. Human rights legislation has been another regulatory avenue used by states in attempts to guarantee inclusive environmental design. In recent years, a number of Western states have enacted various forms of national disability rights legislation with the aim of improving disabled people's access to built environments, and to social life in general. Again, however, there is evidence to show that the rights-based approach to combating discriminatory design has serious political and institutional limitations. In Chapter 9 I will examine the regulation of environmental accessibility in some detail.

Socio-spatial exclusion

In addition to the problem of inaccessibility within public urban spaces, disabled people also experience barriers to choice in their preferred living environment in the contemporary Western city (Dear, 1992; Steinman, 1987). These two areas of socio-spatial injustice present difficult policy challenges, to say the least, for Western governments, most of whom have struggled to lessen the constraints experienced by disabled people in obtaining both employment and a preferred living setting. Not surprisingly, the exclusion of disabled people from employment realms is mirrored in the housing sector. Oliver (1991) argues that disabled people in contemporary British cities suffer housing poverty due both to income deprivation and the discriminatory effects of housing markets that ignore needs for non-standard forms of accommodation. A similar problem has been recognised in Australia (Campbell, 1994; Le Breton, 1985) and in the United States (Dorn, 1994; Harrison and Gilbert, 1992). In 1993, for example, it was estimated that 13,500 disabled Australians had unmet needs for accommodation and respite services (*Canberra Times*, 21 November 1997: 3).

The combined effect of poverty, inaccessibility and inappropriate accommodation is to reduce the ability of disabled people to participate in the mainstreams of urban social life. Gilderbloom and Rosentraub (1990: 271) argue that in many United States cities disabled people 'are often trapped in restrictive living units and are unable to gain access to a city's resources by transportation systems not adapted for them'. For these authors, such cities were no less than 'invisible gaols' for disabled people. Moreover, the powerful gender norms that govern women's embodiment in the mainstreams of city life can mean that disabled women are 'doubly handicapped' in public space (Butler and Bowlby, 1997, Parr, 1997b).

In the late 1980s the United States National Council on Disability undertook a survey of disabled people's lifestyles and came to the following disturbing conclusion:

> The survey results dealing with social life and leisure experiences paint a sobering picture of an *isolated and secluded population of individuals with disabilities.* The large majority of people with disabilities do not go to movies, do not go to the theater, do not go to see musical performances, and do not go to sports events. A substantial minority of persons with disabilities never go to a restaurant, never go to a church or synagogue. The extent of non-participation of individuals with disabilities in social and recreational activities is alarming.
>
> (cited in Harrison and Gilbert, 1992: 18) (emphasis added)

Historically, state support services have been a major cause of the socio-spatial isolation of disabled people. Large institutions have provided both residential 'care' and 'sheltered' employment for disabled people for much of the twentieth century. The oppressive experience of institutionalisation by disabled people was frequently characterised by, *inter alia*, material privation, brutalising and depersonalised forms of 'care', a lack of privacy and individual freedom, and separation from friends and family (Horner, 1994; Shannon and Hovell, 1993).

The failure of institutions as socialised forms of care for disabled people exposes, among other things, the inadequacy of the welfarism which broadly framed the urban social policies of Western states since the Second World War. Institutions may have distributed a very minimal level of material support to disabled people (which admittedly improved in many countries over time), but they also ensured the socio-spatial exclusion of disabled people from the mainstreams of social life, thus entrenching the political invisibility and powerlessness of this social group. In the final section of this chapter I consider Young's (1990) critique of the welfarist model of care and also the potential contribution of her criticisms to an alternative political-ethical ideal, enabling justice.

Recognising the inadequacies of welfarist forms of care, Western governments have sought to deinstitutionalise support for disabled people. This has usually involved both the closure of large-scale residential centres and their replacement with small, dispersed community care units. I will address more fully the limits of this set of policy reforms in the next chapter. Suffice it to say now that many disabled people in most Western countries remain in poor-quality and inappropriate forms of accommodation. In the United States, for example, Dorn reports that 'over two and a half million people with disabilities are warehoused in nursing homes and other institutions at a national cost of approximately [US]$140 billion' (1994: 211).

The social geography of deinstitutionalisation has been thoroughly documented for North America in a set of landmark studies by Michael Dear and Jennifer Wolch. *Landscapes of Despair* (Dear and Wolch, 1987) traced the construction of new urban 'zones of dependence', being clusters of service-dependent groups and facilities designed to support them, usually located in declining inner city areas (see also Dear, 1980, and Joseph and Hall's (1985) examination of clustering in Toronto). Both this, and a follow-up study, *Malign Neglect* (Wolch and Dear, 1993) emphasised how poor public funding and community opposition had forced many deinstitutionalised people into homelessness and 'ghettoisation' in the emerging zones of dependence. Milligan's recent (1996) analysis examined the applicability of these North American findings to Scotland. Milligan concluded that while deinstitutionalised people suffer socio-spatial exclusion in Scotland, this marginalisation departs from the common North American experience due to the mediation of different legislative mechanisms, policy structures and service provision forms.

One major weakness of deinstitutionalisation initiatives has been their lack of congruence with urban planning policies and regulations. In many Western countries, the process of creating community care networks has been slowed, or in some instances actually halted, by planning and building regulations. This dilemma of policy divergence is taken up in the next chapter.

Spaces of resistance

In countries such as the United States, Canada, Britain and Australia, resistance against disability oppression has been rising over the past few decades. Much of this resistance has occurred in cities, and has included frequent and dramatic demonstrations of disabled people's anger and frustration with oppressive structures and institutions. Disabled people have targeted large urban areas in their resistance campaigns, recognising that the city hosts both the mainstreams of public political life in Western countries and also the centre-points of many of the institutions that have contributed to their oppression. Moreover, disabled people have focused their activism on the political city – regional and national capitals – in order to maximise the profile and impact of their campaigns. As Dorn (1994) explains, a common feature of disability activism in the United States has been dramatic seizures of public spaces in and around places such as courthouses, government buildings and public transport systems.

One group in particular, the American Disabled for Accessible Public Transportation (ADAPT), has favoured this spatial politics of resistance, including actions where 'ADAPT activists throw themselves out of their wheelchairs and crawl up the massive stone steps in front of the Capitol Building in Washington (Dorn, 1994: 160). A recent instance of this strategy was the ADAPT protest in early November 1997 outside the White House gates. This action in favour of a national attendant care policy resulted in the arrest of ninety-two activists. Around the same time, ADAPT protesters shut down the Federal Department of Transportation building in Washington for five hours in a dramatic escalation of their struggle for accessibility on inter-city coaches.[10]

Australian disability activists used a similar tactic in July 1997 when they besieged the Prime Minister's Sydney office, protesting against cuts to the budget of the national Human Rights and Equal Opportunities Commission. The protest gained a high profile in the national media, and was described dramatically in one national radio report as 'a stand-off between protesters in wheelchairs and armed guards outside the Prime Minister's office' (ABC Radio News Report, 4 July 1997). In the next month, disabled people in Melbourne staged a protest outside the State Premier's office using their wheelchairs to run over and demolish computers. The demonstration was aimed at new public policies that emphasised the provision of technological aids – especially personal computers – as the answer to disabled people's social needs. As a protest spokesperson argued, the neo-liberal State Government's cuts to basic support services had created a social crisis for disabled people that technological aids could not solve: 'The Internet cannot respond to crisis and people need crisis response because many people out there are in crisis' (ABC Radio News Report, 18 August 1997).

This is not to imply, however, that disability activism has taken the form only of dramatic actions in public space. As a set of (largely urban) social movements, disabled people's organised resistance against oppression has worked at many political levels, including within major political parties. However, the marginalisation of disabled people from mainstreams of power, including formal political spheres, has encouraged the practice of direct action in public spaces. To use Fraser's (1997a) terminology, the disability movements of various Western countries have constituted themselves as 'subaltern counterpublics' that have opposed hegemonic and discriminatory constructions of 'the public sphere'. However, these counterpublics have varied significantly across and within countries – in particular, the tactics used by various national and regional advocacy groups have differed, reflecting specific cultural, institutional and legal contexts.

It is not my intention here to provide a history or detailed contemporary profile of the various national disability movements. Historical accounts of disability social movements have already been written – for example, in the chronicles on the United States by Shapiro (1993) and Britain by Campbell and Oliver (1996). I do, however, think it important here to point briefly to some broad character differences between the national disability movements in the English-speaking world. These differences, both in political approach and in the social gains achieved, reveal the limits of some strategies of resistance to disability oppression. In particular, I believe that the different experiences of disability movements expose the limitations of a rights-based model of resistance, an issue I explore in greater detail in Chapter 9.

As Dorn (1994) and Imrie and Wells (1993a) point out, the disability movements of the United States and Britain have tended to pursue quite different advocacy strategies. In the United States, the disability movement has long followed a militant rights-based course, traceable to the broader eruption of civil rights struggles in the 1960s. By contrast, the British 'disabled people's movement', to use Campbell and Oliver's term, has focused less on the pursuit of individual rights and more on the achievement of social policy gains. The absence of a written constitution laying out individual rights in Britain has lessened the appeal of the rights-based advocacy model in that country (Imrie and Wells, 1993a). Moreover, in Britain, organised charities have played a far greater role in the struggle for progressive disability legislation than has been the case in the United States where advocacy groups have been at the forefront of anti-discrimination struggles (Dorn, 1994).

Imrie and Wells (1993a) claim that in contrast to the United States experience, the British movement has been characterised by political conservatism and social conformism. Some key members of the British disabled people's movement would dispute this depiction – Oliver, for example, is critical of the political-economic cast of the United States movement.[11] In particular, he criticises the latter's focus on the pursuit of individual rights and 'independent living' (IL) for disabled people, which he sees as inferior to a more collectivist approach aimed at changing basic social structures.

> there has always been a distinction between what we mean by IL in Britain and what they meant in the States. IL in America is organised around self-empowerment, individual rights and the idea that in the land of the free and the home of the brave – all that crap – individuals, if they are given access under the law and the constitution, can be independent. In contrast in Britain … IL entailed collective responsibilities for each other and a collective organisation. IL wasn't about individual self-empowerment; it was about individuals helping one another. Once you accept that notion … you are beginning to question the foundations of the society in which we live. It is bizarre for people to think that we, as disabled people can live in Britain with full civil rights and all the services we need without fundamental changes. We are not actually talking about tinkering around at the edges of society to let people in. *For disabled people to play a full part in British society, this society will have to change fundamentally.*
>
> (cited in Campbell and Oliver, 1996: 204) (emphasis added)

In Australia and New Zealand, disability movements have pursued a 'hybrid strategy' that has aimed to secure both improved civil rights and also social structural change, mainly through initiatives in state policy regimes. In Australia, the latter strategy was partly successful during the 1980s, measured in a series of legislative and programmatic initiatives by State and Federal governments that sought to address the employment and income dimensions of disability poverty (Gleeson, 1998). However, at the time of writing (early 1998), these policy gains appeared vulnerable to the cost-cutting agendas of new conservative State and Federal governments.

In most English-speaking countries, there now exists some form of national civil rights legislation protecting disabled people from discrimination, although the strength and effectiveness of these laws varies considerably. The United States Americans with Disabilities Act (1990) is probably the strongest rights legislation. By contrast, the hard-won British Disability Discrimination Act (1995) provides a considerably weaker set of protections for disabled people (Butler and Bowlby, 1997). Australia passed a national Disability Discrimination Act in 1992 (Yeatman, 1996), while in New Zealand disability discrimination was dealt with through the enactment of a Human Rights Act in 1993 (Stewart, 1993). I will return to the issue of rights legislation in Chapter 9.

Urban social movements need, I would argue, a set of political-ethical principles that can guide resistance struggles by supplying both a theory of injustice specific to the group in question and also the criteria for emancipation. This requires both a broad and inclusive ethical ideal, such as an end to all forms of disability oppression, and also a set of subsidiary principles that can be applied to the various spatio-temporal contexts, affinity groups and individual struggles that together constitute the broad phenomenon we denote as a 'social movement'. In the next and final section I outline one ethical principle which might help to focus and evaluate transformative politics in one disabling social domain, the arena of public policy.

Enabling justice

Justice in question

As I have shown in the foregoing discussion, disabled people in contemporary Western cities and societies endure a multifaceted form of social oppression. The obstinacy of this oppression, in spite of the long history of public and private efforts to 'help' disabled people in Western countries, would seem to indict the fundamental political-ethical bases assumed by many of these reformist traditions. Therefore, in order to distil a new political-ethical principle that might aid the emancipatory struggles of disabled people in one important arena, public policy, I think it necessary to begin with a critical look at the broad question of social justice itself. If justice is the antithesis of the heterogeneous oppression facing physically impaired people, then political and institutional remedies must address the full range of needs that disabled people have denied to them in

Western societies. As I will show, recent conventional approaches to justice in Western countries have failed to appreciate both the diversity of disability oppression and the set of deep, interrelated *socio-spatial* changes that are needed to remove this form of disadvantage.

The limits of distributional justice

In her (1990) critique of universalist ideals of justice, Young argues that the social facts of *domination* and *oppression* must replace material distribution as the central politico-theoretical concern for progressive social movements. For Young, established 'welfarist' notions of justice, such as Rawls's (1971) influential formulation, are premised on a misleading social ontology that overlooks the fact of human difference by instating an abstract 'citizen subject' as the beneficiary (or otherwise) of material distributions. As Young shows, the political and institutional practice of distributional justice by capitalist Welfare States in the postwar era was hardly a universally beneficial project, and in fact enshrined the economic and cultural privilege of dominant identities, notably white, middle-class men.

Among the many social groups which the welfarist project has allegedly excluded, Young identifies women, gays, indigenous populations and disabled people as 'marginal identities'. Young notes that marginalised 'groups are not oppressed to the same extent or in the same ways' (1990: 40). Thus, distributive justice is deconstructed as a political principle rooted in the specific social experience of the Western Welfare State. Moreover, Young argues that this highly partial form of justice has in the past depoliticised large areas of political and economic life, including the deep relations that generate injustice (Young, 1990). Her critique of welfarism finds support in much of the critical disability studies literature to emerge in recent years (e.g., Oliver, 1990). Macfarlane, for example, observes that in Britain 'Disabled people have survived in a society which, historically, has not included them in any sort of consultation. This has resulted in totally inappropriate forms of service provision' (1996: 7). Clearly, institutional 'care' is one such (dis)service endured by generations of disabled people. Incarceration in remote institutional settings was (and remains) a core feature of the injustice of cultural exclusion and invisibility that disabled people have long experienced.

Young's critique of the Welfare State seeks a constructive engagement with the Rawlsian distributive paradigm, attempting to transcend this theory of justice, rather than dispense with it altogether. Her argument is that this conception of social justice was limited both by its focus on distributional outcomes in society and by an unwillingness to consider the unequal power relations which generate them. Similarly, Fraser (1995: 84) describes the redistributional policies of Welfare States as 'affirmative remedies' that 'seek to redress end-state maldistribution, while leaving intact much of the underlying political-economic structure'. At best, welfarism merely compensates disabled people for their oppression through, *inter alia*, transfer payments, various social support mechanisms and therapeutic programmes. However, such 'affirmative remedies' cannot address the causes of oppression for disabled people, and may actually worsen their disadvantage by codifying the relationship between disability and social dependency.

Young also argues that the distributive principle is too narrow, addressing only the fair allocation of *material* resources in society. She argues for an expanded ontology of needs which embraces more than material necessities, such as food and shelter, and includes fundamental human desires for social participation and freedom from oppression on the basis of shared qualities such as race, gender and (dis)ability. However, Young's critique of the 'distributive paradigm' is itself narrowly based and tends to caricature the distributional justice perspective. Rawls (1993) has recognised the importance of cultural issues to justice, though not within a 'politics of difference' framework. As Fraser (1995: 71n) points out, 'Rawls ... treats the "social bases of self-respect" as a primary good to be fairly distributed'. More broadly, Young's appraisal overlooks the considerable critical attention which several key advocates of distributional justice have given both to problems of institutional arrangements and to issues of social inclusion. A few examples will substantiate the point.

First, the influential work of Doyal and Gough (1991) can be set within the distributional paradigm, given the insistence in their analyses upon centrally planned systems of need satisfaction. Doyal and Gough's support for institutionalised systems of need satisfaction is highly nuanced and critical of practices in contemporary Welfare States. They stress that 'central planning *and* democratic participation are both necessary components of social policy formation if it is to succeed in optimising need satisfaction' (1991: 297) (original emphasis). Similarly, McConnell (1981), a strong advocate of distributional justice in urban planning, criticised Rawls's naive consensualist view of society and argued that the institutional systems of most Western democracies had excluded many disadvantaged social groups. Moreover, the geographer Badcock (1984) provided a comprehensive analysis of urban inequality and argued for a (re)distributional justice which would overcome the limitations of the welfarist approach. In short, while Young's criticisms of the Welfare State have undoubted veracity, she has clearly overlooked the considerable critical appreciation of these institutional failings by theorists identified with the distributional paradigm.

Fraser (1995, 1997a) has also provided a thoroughgoing analysis of social justice thought that criticises the distributive paradigm for its inattention to cultural issues while also pointing to the political dangers inherent in the new turn to a 'politics of difference'. Fraser recognises that, in some circumstances, the stress on social difference may elide both the ideals of human rights and social equality (as, for instance, in political sentiments based upon national or ethnic affinities). Therefore, in arguing for a new distributive paradigm that stresses both material and cultural justice, Fraser is prepared to accept 'only those versions of the cultural politics of difference that can be coherently combined with the social politics of equality' (1995: 69).

To some extent, Young's critique of the distributive paradigm draws upon postmodernist and feminist criticisms of established formulations of social justice. Postmodernists have identified a critical set of limits for welfarist justice by arguing that such ethical principles cannot be applied meaningfully, or at least not in any benign way, outside the spatio-temporal context of Western capitalism. Moreover, feminists, some with postmodern inclinations, have criticised mainstream theories of social justice which enshrine impartiality within the public sphere (Hekman, 1995; Smith, 1994). Among these, Gilligan (1982) and Tronto (1987, 1993) have argued for an alternative 'ethic of care' which recognises the tendency of people – especially women – to favour 'local and familiar' contexts in their everyday moral decision making. Thus feminists have identified two further critical limitations of putatively universal theories of justice: the erroneous reduction of society to the public sphere; and the exclusion of affective ethical values, such as care, from moral consideration.

Many objections have been raised to the idea that an 'ethic of care' can supplant social justice as a political principle for society. First, it is not clear how affective relations can provide a moral basis for decision making in the non-domestic spheres (e.g., paid work, public and private institutions) where women are increasingly involved as formal participants. Moreover, the sources of injustice are rarely locally specific, with whole societies, even the globe, subject to structural sources of disadvantage, such as economic exploitation and racism (Mendus, 1993). Other commentators have pointed to the potential for an ethic of care, in the absence of universal political standards, to justify cultural parochialism and, ultimately, even the horrors of ethnic and racial conflict (Tronto, 1987, 1993).

Harvey (1993) has offered a measured response to the postmodernist critique of universals, such as social justice, drawing upon Young's (1990) work. Harvey opposes the (de-politicising) prospect of moral relativism while also recognising the limits of many modernist ethical formulations which have marginalised 'others' by ignoring the critical fact of human social difference. Harvey doubtless agrees with feminist philosophers, such as Kearns (1983) and Pateman (1980), who have argued that mainstream theories of justice have in the past simply assumed away the family as a potential site for injustice, thereby excluding many women from the 'meta-ethical' purview.

Notwithstanding their various theoretical differences, it can be said that Young (1990), Fraser (1995, 1997a) and Harvey (1993, 1996) all support the idea of an inclusive notion of justice which draws upon a range of social affiliations and viewpoints. This approach to political ethics would avoid the tendency of certain modernist approaches to *impose* a general moral outlook. Importantly, all three observers disavow *a priori* ethical formulations, and their prescriptions of justice are sourced in careful appreciations of concrete forms of oppression and social exclusion.

Enabling justice

Broadly then, a new formulation of social justice can be identified which combines the arguments of Fraser (1995) and Young (1990) with various critical antecedents in the distributive paradigm. As Fraser (1995: 69) puts it, this new critical social justice must combine the ideals of 'cultural recognition and social equality in forms that support rather than undermine one another'. Fraser and Young both argue that justice can only obtain if individuals and groups are enabled to participate in the mainstreams of social life in meaningful ways. Clearly, 'enablement' means more than the satisfaction of material needs; cultural empowerment is acknowledged as an equally necessary condition for social participation. I therefore propose the term 'enabling justice' for this new ethical formulation, in order to reflect the twin emphases upon material redistribution and cultural recognition as mutually dependent political ideals.

Specifically, this new distributive justice must uphold the right of all to have their material needs guaranteed, while insisting on the necessity of freedom from cultural-political forms of disability oppression (cf. Wilmot, 1997: 44–72). Enabling justice centres upon a socially codified guarantee that all individuals and collectivities are entitled to have their basic needs fulfilled. Importantly, these needs are seen to have two dimensions:

- material satisfaction (i.e., minimum access to food, shelter, personal items, etc. defined by average social consumption patterns), and
- socio-cultural participation (i.e., affective and social ties, political inclusion, cultural respect).

These justice criteria do not seek 'independence' for disabled people, at least not in the sense imagined by neo-liberal individualism. Rather, enabling justice stresses a *social* ontology, and from this, the goal of mutual interdependence for all people and affinity groups. This position reflects the long-established radical critique of ideologies of individualism and self-reliance that have been used to sustain unequal power relations in capitalism (Jary and Jary, 1991). It also embodies the more recent criticisms that disabled commentators have levelled against the ideal of independence for service dependent people that has been promoted by neo-liberal states anxious to reduce the loads on national welfare budgets (French, 1993b). Oliver (1993) makes the point that disabled people want social inclusion and cultural respect rather than individual independence, a goal that recognises the inescapable fact that all agents are constituted through, and dependent upon, networks of mutuality that range in scale from the local-affective to the national-institutional. The real question is what types of (inter)dependence are consistent with enabling justice? Certainly, the economic dependence of disabled people on state services – a historical product of both market forces and public policies – does not meet the demands of justice.

Finally, the two 'tests of justice' listed above embody the bivalent remedies which Fraser (1997a) sees as necessary to the emancipation of groups subject to multiple forms of oppression. This point is supported by Morrison and Finkelstein (1993) who articulate the crucial role that cultural empowerment must play in the struggle against disability oppression.

Of course, there are qualifications to be made. In particular, the geographical dimensions of justice must be considered. Justice, like any other political ideal, must be realised within the context of socio-spatial relations. What then is the environment of enabling justice?

Towards enabling environments

Radical geography and justice

During the 1970s, advocates of the (now largely dormant) welfare perspective applied the question of social justice to geographical analysis (e.g., Smith, 1977). This new emphasis on social equity represented an important break from the prevailing positivism of Human Geography (Johnston *et al.*, 1994). In this sense, Welfare Geography represented an important early 'radical' impulse in Social Geography. However, there were significant conceptual and political limits to Welfare Geography and the perspective has been criticised for its tendency to focus only upon the *distributional consequences* of the material and ideological structures that condition the production of space (Badcock, 1984; Johnston *et al.*, 1994). This is to echo the criticisms of Fraser (1995) and Young (1990) of the conceptual and political limits of welfarist notions of equity.

By contrast, the subsequent tradition of radical Social Geography – in particular, historical-geographical materialism – provides a more promising conceptual basis from which to formulate a spatial notion of enabling justice. The key insight of radical social geographical analysis in this respect is the view that oppression and exclusion arise from the socio-cultural production of space. Recalling the analysis of Chapter 3, this ontology sees society and space as mutually constitutive dynamics. Importantly, 'environment', as the physical and social context of life, is assumed to be an artefact of human society, rather than merely a surface upon which materialities are rearranged. This spatial ontology thus problematises the justice (or otherwise) of structures which *produce* space in capitalist societies. As Young (1990) points out, the capitalist city is an environment where injustice has been produced in multiple, interdependent forms. A radical enabling justice would thus presuppose the broad ethical and political goal of producing environments which liberate the social capacities of all people. Put differently, enabling justice requires the production of spaces and places which guarantee the capacity of all to participate in social life in meaningful ways, such that each individual's material and non-material needs are satisfied (e.g., inclusion, affectivity, liberty).

While the ethic of care cannot replace the need for meta-ethical formulations, the criticisms which many feminists have levelled against de-contextualised notions of justice should be borne in mind. An enabling justice would recognise that the universal need for material welfare, social participation and cultural respect must be realised at a variety of socio-spatial scales, each defined by unique sets of affective and social ties, group affiliations, and environmental conditions. In this sense, an ethic of care complements rather than contradicts justice by stressing the need for contextually appropriate, rather than uniform, mechanisms for material distribution and social participation.

Ethical contextualisation has also been an important theme of communitarian theorists, several of whom are implacably opposed to universalist notions of justice (Smith, 1994). Communitarians such as Sandel (1982) and Walzer (1983) have stressed the 'community' (defined at a variety of socio-spatial scales) both as the source of unique, context-bound moral frameworks and as the most appropriate sphere for ethical practice. In recent years, a diluted form of communitarian thinking has played an influential role in political discourses within a range of Western (mostly English-speaking) countries. For example, parties and thinkers from both the Right and the Left have promoted versions of 'welfare pluralism' which, while differing on some key political-economic grounds (e.g., the extent of marketisation of the public sphere), none the less share an emphasis on the community as a vehicle for decentralised, participatory structures of social service delivery (Clapham and Kintrea, 1992; Jary and Jary, 1991).

Neo-liberals, in particular, have championed 'the community' as a fulcrum of moral responsibility and efficient social adaptability in contradistinction to the supposed inflexibility and unaccountability of state institutions. However, it can be argued that the real agenda of neo-liberal communitarianism is to shift the costs of morally based action, such as the provision of social support services, from the state to local communities and individuals (especially women) (Jary and Jary, 1991). Given the uneven capacity of communities and individuals to resource social support, this reallocation of costs and service responsibilities inevitably worsens distributional injustice. Although speaking the language of contextualised ethics, this form of communitarianism is clearly antithetical to enabling justice.

Transforming the environment of disability

What would enabling justice mean for disabled people? At the minimum, the goal of enablement demands the creation of new social spaces that 'accommodate a broader range of human capabilities than the present environment' (Hahn, 1987b: 188). Thus, disability scholars and activists have called for the creation of *enabling environments* in capitalist societies which emphasise the capabilities rather than the impairments of disabled people (see, for example, the collections edited by Hales, 1996 and Swain *et al.*, 1993). For Corker (1993), the 'enabling environment' would aim to establish social independence for all inhabitants, meaning that disabled people, in particular, would be empowered to meet their own needs within a network of mutual obligations rather than within a hierarchy of dependency relationships (e.g., care giver/care receiver). Finkelstein and Stuart echo this theme, envisaging the socio-cultural emancipation of disabled people through a wholesale transformation of public policies:

> In [this] new world ... services for disabled people should be conceived in terms of 'support' and would acquire an *enabling* role in the same way that public utilities (for example, postal services, railways, water and electricity supplies, and so on) are created by able-bodied for able-bodied people to enable more satisfying life-styles. As such, they form part of the necessary

public support network which *enables* both full participation in society and citizenship rights.

(1996: 171) (emphasis added)

The complementary ideals of full citizenship rights and social independence would require the integration of disabled people within both mainstream political settings and principal economic spheres, especially labour markets (cf. Kavka, 1992).

As Harvey (1996) argues, broad social change is realised through the multiple forms of spatial struggle that attempt to create material, representational and symbolic places of emancipation. Thus, the 'enabling environment' might range in scale from the level of a local policy sphere which empowers disabled people to meet specific needs (e.g., accommodation, education, work) to that of a whole society which has ceased to oppress and exclude people on the basis of any social difference. Although possessing shifting, indeed contested, geographies, the disabled people's movements are good examples of the specific enabling environments – or 'subaltern counterpublics' – that have arisen within cities in opposition to structures of oppression. If generalised to the level of society as a whole, the enabling environment ideal would restore to disabled people the material needs, cultural respect and political voice that many are at present denied.

Thus the broad definition of enabling justice offered earlier can be re-stated in more specific terms for disabled people as:

- the satisfaction of material needs, as socially defined in the relevant regional or national context;
- socio-political participation and cultural respect; and
- socio-spatial inclusion.

It may be objected that these conditions for justice are utterly quixotic given that a defining feature of capitalism – commodity relations – has been implicated in the economic devalorisation, and therefore social oppression, of disabled people. The first response to this criticism – versions of which are commonly directed at all 'radical' political movements – is that many disabled people themselves have insisted that nothing less than profound socio-spatial change will remedy the oppression they endure. Finger, for example, insists that the realisation of enabling work environments requires a fundamental transformation of capitalist labour markets, involving, *inter alia*, replacing the law of value with a new social measurement of economic usefulness:

we need to argue against 'productivity' and 'bringing home a paycheck' as a measure of human value. We need to work for a society that values a range of kinds of labor and ways of working – everything from raising children to working for disability rights.

(1995: 15)

The justice criteria listed above reflect the emancipatory demands levelled by disabled people in Western countries through both advocacy fora (e.g., Disability Alliance, 1987a, 1987b; Eastern Bay of Plenty People First Committee, 1993; Ronalds, 1990; UPIAS, 1976) and theoretical discourses (e.g., Abberley, 1991a, 1991b; Morris, 1991; Oliver, 1990, 1996; Swain *et al.*, 1993).

However, this answer does not respond adequately to the charge of utopianism that conventional critics always level at transformative principles, such as enabling justice. Indeed, given the obvious difficulties of removing key sources of disability oppression in an era of 'market triumphalism', what practical political purpose can the principle offer to radical social scientists and activists?

My argument is that enabling justice – and the subsidiary ideal of an enabling environment – can provide the basis for progressive political practice in state policy spheres. Obviously, these broad ideals have implications outside public policy realms – however, the emphases on material distributions, spatial inclusion and citizenship have a particularly strong resonance in the state arena. Moreover, some commentators have argued that the state must take a lead role in countering disability oppression through the enactment of enabling policies and legislation. Oliver and Barnes (1993: 275), for example, argue that the state must 'cease its current discriminatory welfare provision and move towards forms of provision which are truly enabling'. Importantly, the enabling principle problematises the attempts of conventional public policy frameworks to address the causes and outcomes of social oppression. In particular, the principle of enabling justice can be used to interrogate, and – it is to be hoped – exact policy concessions from, state institutional practices which affect the well-being of socially oppressed people. I intend to demonstrate the value of the principle in the two chapters which now follow.

Conclusion

This chapter considered the experience of disablement in the contemporary capitalist city. My review cast disablement as a distinct form of social oppression whose genesis arises from broad socio-cultural arrangement of capitalist societies. 'Disability oppression' was explained, in Fraser's terms, as a bivalent form of disadvantage, embodying both political economic hardship and cultural misrecognition. Several key dimensions of disability oppression were examined, including labour market exclusion, poverty, socio-cultural devaluation, and socio-spatial marginalisation.

As seen in Chapter 6, the situation of most physically impaired people in the nineteenth-century industrial city was frequently wretched, and resistance to the social oppression of disablement was difficult, if not completely impossible. While material conditions might now be better for disabled people in Western societies, the contemporary city none the less continues to oppress impaired embodiment through both its socio-structural organisation and its physical layout. However, in contrast to the past, disabled people have in recent years organised themselves into various national and regional social movements that have openly resisted disability oppression. This resistance has largely, though not exclusively, occurred in large cities.

In the final part of the chapter I sketched in outline a political-ethical principle that demands enabling justice for all disabled people. The ideal focuses on the questions of citizenship and socio-spatial inclusion, matters of central relevance to the responsibilities and activities of capitalist states. To varying degrees, all states have had to respond to the campaigns waged by disabled people's social movements against disability oppression. As was explained, many states have responded to this socio-political pressure by enhancing the human rights protections for disabled people through specific legislation. Other responses have included various policy concessions and initiatives that putatively remove or reduce the disadvantage that disabled people experience in various state and civil arenas. However, many of these state legislative and policy responses have been criticised, at times bitterly, by disabled people as unhelpful and even retrograde. In the next two chapters, I will consider in turn two broad initiatives in disability policy that have been embarked upon by most Western states in recent decades: deinstitutionalisation and accessibility regulation. In both instances, my analysis of policy practice will measure the extent to which these initiatives have met the demands of enabling justice.

8 Community care: the environment of justice?

Introduction

This chapter will examine one area of state policy practice in advanced capitalist societies – the establishment of community care networks for socially dependent persons. In this discussion, 'community care' refers to the 'care of individuals within the community as an alternative to institutional or long-stay residential care' (Jary and Jary, 1991: 99), a service principle that is now a common feature of state social policy in advanced capitalist nations (Heginbotham, 1990; Lerman, 1981; Mangen, 1985; Prior, 1993). In most Western countries, community care has been achieved through a programmatic deinstitutionalisation of social support, involving the closure and/or downscaling of large-scale human service facilities (Bean, 1988; Kemp, 1993; Smith and Giggs, 1988). Community care has sought to improve the well-being of publicly dependent disabled people by providing for their support in dispersed, small-scale residential settings (Parker, 1993). The policy of community care, as explained by advocates (e.g., Lakin and Bruininks, 1985; Perske and Perske, 1980) is essentially the *humanisation* of established modes of social support. The policy is argued to improve the quality of social care, reduce the restrictions on individual liberties which were a feature of large-scale institutions, and promote the re-integration of dependent peoples into the broader community.

One consequence of Welfare State restructuring since the 1970s (especially in Anglophonic countries) is that such 'care' is increasingly provided by a diverse, non-government human service sector, made up of a panoply of voluntary and for-profit agencies (Smith and Lipsky, 1993; Wolch, 1990). According to Smith and Lipsky (1993), any 'contracting out' of human services by the state is a form of privatisation, irrespective of whether the supplier is motivated by profit or by altruism. In Britain, for example, this localised, multi-provider approach to community care was enthusiastically described as 'welfare pluralism' by the influential 1989 White Paper, *Caring for People: Community Care in the Next Decade and Beyond* (Jary and Jary, 1991). Governments in Australia and New Zealand have also favoured the multi-provider model in recent years (Lyons, 1995; Fougere, 1994).

While the policy of community care alone cannot deliver enabling justice for disabled people, it may well lessen one dimension of the oppression which they experience: their socio-spatial exclusion in remote, often de-humanising, institutional settings. I therefore begin my analysis in this chapter from the premise that community care is a potentially enabling public policy initiative (cf. Doyal, 1993). My aim here is to measure the *socio-political* potential for the realisation of community care networks in Western countries where the policy has been established.[1] This socio-political perspective can be explained as follows.

Community care represents the state's attempt to produce a landscape of dependency that is superior to antecedent forms (institutional environments) in terms of presumed human rights to freedom and social participation. This prospect is dependent upon the successful establishment of residential support programmes within mainstream living settings. However, the programmes must be realised within an overall production of social space that is conditioned by a range of structural and institutional dynamics, including capitalist commodity relations, state ideologies and practices, and socio-cultural attitudes within civil society itself. These broader processes thus condition the potential for human service agencies and state bodies to produce the social space of community care.

In this chapter I will assess the effect of these broader conditioning processes on the policy practice of community care. This assessment will consider the relative success of this policy practice in reducing the injustice that many disabled people have experienced within institutionalised landscapes of 'care'. As I will show, an examination of the practice of community care in particular national contexts reveals several problems which may frustrate the realisation of its policy aims, including opposition to care facilities from nearby residents; the inability of planning systems to accommodate social policy concerns; and structural changes to social policy by neo-liberal governments.

The chapter is structured as follows. In the first section I review the process of deinstitutionalisation that has generated the new landscapes of community care. The next three sections examine in turn the major socio-political threats to community care in the form of locational conflict, inadequate planning systems and welfare state restructuring. The chapter concludes by reflecting on the capacity of community care policies to provide enabling environments for disabled people in advanced capitalist societies.

Deinstitutionalisation

Most Western countries are currently reducing the scale of, and in many cases completely closing, large institutions that have provided both residential 'care' and 'sheltered' employment for disabled people for much of the twentieth century (Pinch, 1997; Shannon and Hovell, 1993). The oppressive experience of institutionalisation by disabled people was frequently characterised by, *inter alia*, material privation, brutalising and depersonalised forms of 'care', dangerous and/ or unhealthy living conditions, a lack of privacy and individual freedom, and separation from friends and family (Dear and Wolch, 1987; Horner, 1994). Decades of reforms and improvements to institutions have not eliminated these inhumane conditions and practices – across Western countries, outrages still emerge with depressing frequency.

In Australia, for example, an inquiry in 1996 by the New South Wales State Government into residential care for intellectually and physically disabled people found evidence of an entrenched 'culture of abuse' in both public and private institutions (*Sydney Morning Herald*, 30 November 1996: 3). The report detailed harrowing accounts of sexual and physical abuse of residents, both by staff and by fellow-residents.[2] In April of the same year, the perilous nature of Australia's institutions was thrown into ghastly relief by a fire in one ageing public facility in Melbourne that killed nine intellectually disabled men (*The Australian*, 10 April 1996: 1). In a later official inquiry into the fire, the parents' association representing disabled residents of the institution attributed the disaster to the facility's decrepit fire-safety system and the poorly trained and under-equipped staff. At the same time, the State of Victoria's Office of the Public Advocate found that the government had neglected its duty to provide safe residential services for disabled people. A later investigation by a major Melbourne newspaper found similar defects in the State's other major institutions for disabled people (*The Age*, 13 May 1996: 1). A final condemnation of the government's institutional management was made in 1997 by the State Coroner, who found that a decade of official neglect of public facilities for disabled people had contributed to the deaths of the nine men (*Canberra Times*, 18 October 1997: 4).

The failure of institutions as socialised forms of care for disabled people exposes, among other things, the inadequacy of the welfare ideal of justice. Institutions may have distributed a very minimal level of material support to disabled people (which admittedly improved in many countries over time), but they also ensured the socio-spatial exclusion of disabled people from the mainstreams of social life, thus entrenching their political invisibility and powerlessness. As Oliver and Barnes observe on this point:

> not only has state welfare not ensured the basic human rights of disabled people, through some of its provisions and practices it has infringed and even taken away some of these rights. Examples of this include the provision of segregated residential facilities which deny some disabled people the right to live where they choose.
>
> (1993: 267–8)

Accordingly, Western governments have sought to replace institutional support for disabled people with community care networks (Bennie, 1993; Lerman, 1981; Prior, 1993). Community care is usually undertaken in networks of small, dispersed residential units which aim to provide 'socially valorised' living settings for service dependent peoples (Wolfensberger, 1987). The impetus for reform has come from a number of sources: partly from an acknowledgement by states of the inadequacies of institutional care (both from therapeutic and fiscal viewpoints); and largely from the political pressure applied to governments by various national and regional disability social movements since the 1960s. In the

United States, for example, the shift to community programmes has been hastened by court decisions that have ruled institutions unfit for the purposes of care (Gilderbloom and Rosentraub, 1990).

Deinstitutionalisation has been promoted by advocates in social justice terms as a restoration to service dependent peoples of their basic human right to a valued living environment (Oliver and Barnes, 1993; Shannon and Hovell, 1993; Wilmot, 1997). For service users, deinstitutionalisation has promised the right to the 'least restrictive living setting', meaning a care environment that restricts the freedom of disabled people only to the minimum extent needed to ensure broader community well-being (Shannon and Hovell, 1993). In this sense, proponents of deinstitutionalisation claim that the policy addresses the inadequacies of welfarism for disabled people by radically changing the socio-spatial arrangements of social care, involving the re-integration of service dependent people within the mainstream living environments of Western societies (Swain *et al.*, 1993).

However, the policy experience of several Western countries suggests that community care programmes have been obstructed by community hostility, bureaucratic uncertainty, and fiscal conservatism (Dear, 1992; Grob, 1995). In particular, deinstitutionalisation has faced barriers at the local policy level in the forms of community opposition to residential social programmes and (often ambiguously defined) urban planning controls. In addition, at the structural level, recent changes to social service delivery wrought by neo-liberal governments mean that the therapeutic and human rights ideals which underscore community care may be overshadowed, if not erased, by the logic of profit (Glendinning, 1991; Oliver and Barnes, 1993). These obstructions are now briefly reviewed with reference to the policy experiences of a range of Western countries.

The NIMBY threat

A critical social dynamic has limited the ability of deinstitutionalisation to secure justice for disabled people: pervasive opposition to community care facilities, in the form of the NIMBY ('Not-in-My-Backyard') syndrome (Dear, 1992). Young (1990) identifies localised opposition to community care homes as a major source of injustice for disabled people.

Much of the early investigation of NIMBY reactions to care homes was undertaken in North America where deinstitutionalisation has been under way for at least three decades. The pioneering work of Wolpert (e.g., 1976), Dear (e.g., 1977, 1981, 1992) and collaborators (e.g., Dear *et al.* 1977; Dear *et al.* 1980; Dear and Taylor, 1982), and Smith (e.g., 1981, 1984, 1989) was important both in charting the course of deinstitutionalisation policies and in developing a critical geographical analysis of the NIMBY syndrome as a particular form of urban locational conflict. In recent years both the NIMBY phenomenon and scholarly interest in it have grown within a range of Western countries, including the United Kingdom (e.g., Burnett and Moon, 1983; Locker *et al.* 1979; Moon, 1988), Australia (e.g., Foreman and Andrews, 1988) and New Zealand (e.g., Gleeson *et al.* 1995; Shannon and Hovell, 1993).

This geographic analysis of NIMBY and locational conflict has mostly centred on reactions to facilities for people with mental illnesses. I contend that this literature may not fully appreciate the singularity of locational conflict issues that surround disabled people and the support facilities provided for them. While many of the broad findings of these locational conflict studies are relevant to the question of disability, there is a need for further analyses that can identify the distinct patterns of community receptiveness towards physical impairment and mental illness. In most Western countries the distinctiveness of these issues is reflected in separate, if sometimes overlapping, policy regimes for disability support and psychiatric care. I will return to this issue in the concluding chapter.

As Dear and Taylor (1982) have shown, the NIMBY syndrome reflects a complex mixture of popular anxieties about various categories of service dependent peoples. Each instance of NIMBY reaction is therefore likely to reflect a specific combination of local community fears about the particular client group for which the facility is intended. There is at least one common factor of concern that arises in most NIMBY disputes: property values. NIMBY sentiments are partly sourced in the deep property interests which structure the capitalist land economy. This fact helps to explain the considerable role that locational conflict has played in impeding the transition to community care (Gleeson and Memon, 1994). As Walker (1981) explains, the commodified nature of residential land in capitalist societies is a powerful influence on homeowners' (and home-purchasers') social interests. For the homeowner, the commodity land has a dual character as both use value (the residential living setting) and exchange value (potential sale price). The exchange value of a homeowner's property measures the worth of the major wealth asset for most households. As a (or *the*) major wealth asset for many households, the home serves an important dual role both as a repository of capital investment and stored equity, and as a source of profit (capital gain). The exchange value of the home, therefore, is a critical influence upon the social interests of many households in capitalist societies.

Locational conflicts are often expressed as defensive reactions by homeowners confronted with land uses that are perceived as threats to residential amenity (the putative 'character' or 'quality' of a residential environment). In fact, the notion of 'residential amenity' is heavily coded with concerns for land as a commodity which is capable both of storing value and rendering capital gain (profit) (Walker, 1981). Seen from this perspective, homeowners' sensitivity towards amenity, and land uses which may threaten this, is an outward expression of their deeper social interests as commodity purchasers and owners who are concerned to safeguard the exchange value of their principal capital possession, residential land. Hence, NIMBY sentiments are often the phenomenal form of deeply embedded class interests; namely, the concern of homeowners to safeguard the exchange value of their principal economic asset (Walker, 1981; Plotkin, 1987).

This would suggest that areas with high concentrations of homeowners, as opposed to other land uses and other residential tenure types, would be most likely to resist the 'intrusion' of perceived noxious facilities (Dear and Taylor, 1982; Dear, 1992; Plotkin, 1987). Such resistance will take the form of NIMBY reactions in which homeowners pursue an important common interest – the protection of residential exchange values – through collective actions that curb the entry of unwanted facilities into their neighbourhoods (Beamish, 1981). These collective NIMBY actions are most often pursued through local government development control systems (Burnett and Moon, 1983; Moon, 1988; Locker *et al.*, 1979).

Commentators in a range of countries – including the USA (Dear and Wolch, 1987), Canada (Joseph and Hall, 1985; Taylor, 1988), Australia (Gleeson, 1996c) and New Zealand (Shannon and Hovell, 1993) – have argued that NIMBY reactions threaten the entire process of deinstitutionalisation by creating (often bitter) political and legal resistance to the establishment of care homes. Dear and Wolch (1987) have shown that North American service agencies have responded to the prevalence of NIMBY by adopting 'avoidance strategies' as part of their siting criteria for community care homes. The outcome of avoidance strategies has been the concentration – or 'ghettoisation' – of care networks in 'places of least resistance', frequently low income and declining inner city neighbourhoods. The strategy of avoidance has also been identified among service agencies in New Zealand (Gleeson *et al.*, 1995).

A series of recent court rulings in Western countries has illustrated the enduring political-economic potency of NIMBY sentiments and their capacity to constrain disabled people's choice of living environment. In September 1995 the British High Court ruled that a set of neighbours were entitled to compensation for a fall in property values after a local health authority had established a care home in their immediate vicinity. The broader implications of this ruling were not immediately clear, but health authorities feared that the fiscal impacts of the decision would jeopardise all community care programmes in the United Kingdom (*The Times*, 21 September 1995: 2). Lamentably, the British court ruling seems to ignore a substantial body of geographic (e.g., Dear and Taylor, 1982; Wolpert, 1978) and other social scientific (e.g., Consulting Group, 1992) evidence from a variety of countries which shows that care homes tend not to affect residential property values in the medium to long term.

Also in 1995, the United States Supreme Court ruled that cities may not use local zoning ordinances to exclude group homes for disabled people from residential areas. The court ruled that exclusionary zoning violated the rights of persons, including disabled people, protected under the anti-discrimination provisions of the federal housing laws (*AAMR News and Notes*, May/June, 1995: 1). However, Zipple and Anzer (1994) earlier reported how city authorities had anticipated such a ruling by switching to other regulatory modes – notably building codes – in order to achieve exclusionary zoning outcomes. On the basis of the US experience, it seems difficult to conclude that mainstream human rights legislation is sufficient to protect disabled people against NIMBY discrimination.

The NIMBY syndrome certainly exercises a pre-emptive power over the process of community care which constrains and distorts the locational freedom of service agencies. In many instances, agencies feel that they lack the power and expertise to confront NIMBY hostility, and their resort to avoidance strategies means that many service dependent disabled people are excluded from numerous contexts for community living. NIMBY actions thus threaten both the therapeutic and the human rights ideals of community care. If community care offers a modest, but important, strategy for reducing the environmental injustice that many disabled people suffer, NIMBY sentiments represent the sort of intolerance to difference which Young (1990) identifies as a defining feature of social oppression.

Planning context

Urban planning and NIMBY

Local community hostility towards community care facilities is often galvanised by bureaucratic passivity or confusion at all levels of government. In particular, the frequently poor articulation of the social and land use policy functions of local governments in Anglophonic countries (de Neufville, 1981) has meant that planning has often been a regressive tool of privileged community interests (Forester, 1989). The lack of a clear social agenda for land use planning has often exposed this mode of local state practice to 'policy capture' by powerful interests in local land economies. Such vested interests, including homeowners, are frequently concerned to shape land use regulations that will safeguard their economic interests by excluding any social groups or facilities which might threaten amenity and/or land values (Plotkin, 1987).

NIMBY reactions are commonly articulated within the regulatory frameworks of local government planning, most frequently as attempts by hostile communities to achieve exclusionary zoning rulings which prohibit care homes on the basis that they are not 'legitimate' residential uses (Benjamin, 1981; Dear and Laws, 1986). The local state planner is inevitably embroiled in such NIMBY disputes, having to arbitrate in the complex social conflicts that often attend the siting of community care facilities (Jaffe and Smith, 1986). Planners thus have a critical role to play in the establishment of community care networks, though this fact is rarely acknowledged in the urban planning and social policy literatures.

As Dear (1992) has explained, the NIMBY syndrome expresses more than simply a concern for property interests. The syndrome reflects deep-seated and intricate fears in the popular mind concerning both the personal (especially behavioural) attributes of service dependent persons and the nature of service programmes and facilities which support such people. Recalling the analysis in the previous chapter, the NIMBY mind-set is one powerful expression of the disabling imaginary, an anxiety about 'unruly bodies' that do not correspond to dominant constructions of safe and desirable forms of embodiment. Seen in this light, the NIMBY phenomenon emerges as a powerful cultural-material force that has helped to reinforce the disabling socialisation of embodiment in recent and contemporary Western societies.

Dear and Taylor's (1982) exhaustive study of NIMBY conflicts in Toronto demonstrated that community fears can centre variously on the service facility, service users, service programmes, or a combination of some or all of these. This and other (e.g., Freudenberg, 1984; Mowrey and Redmond, 1993; Plotkin, 1987) in-depth studies of NIMBY behaviour have drawn upon a range of theoretical frameworks from within social psychology, sociology, political economy and philosophy (the list is not exhaustive) to explain why 'host' communities often fear certain types of land uses.

Planning and social policy

Given the complexity of motivations behind NIMBY sentiments, there is a need for urban planners in Western countries to appreciate both the nature of community care and the questions of justice and human rights which this social policy raises and addresses. To this end, Young (1990) has exhorted planning agencies to adopt regulatory practices that foster diversity and eliminate the socio-spatial exclusion of social groups. For her, 'zoning regulations that limit ... location choices' represent institutional sources of injustice for disabled people (Young, 1990: 255). Moreover, this integrative aim should become a strategic goal of urban and regional planning rather simply a principle applied at the local level of development control:

> regional planning decisions should be aimed at minimizing segregation and functionalization, and fostering a diversity of groups and activities alongside of and interspersed with one another.
>
> (Young 1990: 255)

Are planners trained to understand the complexity and breadth of popular anxieties about service dependent peoples and the facilities which support them? Recent research in New Zealand (Gleeson and Memon, 1994, 1997; Gleeson *et al.*, 1995) suggests that planners in that country are largely both unaware of the policy context for deinstitutionalisation and uncertain about the nature of NIMBY conflicts involving community care facilities. This finding is doubtless not unique to New Zealand given that formal planning education in most Western countries rarely includes exposure to social theory and policy (McLoughlin, 1994). Certainly, in both Britain and British Commonwealth countries, such training would seem vocationally irrelevant given that the separation of social policy and development control functions in local government is a key feature of the British 'town and country' planning system (Cullingworth, 1985). (De Neufville (1981) has made a similar assessment of the US planning system.) Kiernan (1983), reflecting on experience working within the Canadian local planning system, concluded that planners in that country denied both the political nature of their work and its implications for social justice. Indeed,

> since this apolitical planning ideology implicitly denies any negative impact that planning might have on social problems, it must also be oblivious to the potential for planning to intervene in a conscious positive way

to alleviate them. This conception thus dooms planners to a role that is at best irrelevant to the process of social change, *and at worst actively pernicious.*

(Kiernan, 1983: 74) (emphasis added)

Can planning actually harm the interests of the socially vulnerable as Kiernan (1983) suggests? In the case of NIMBY conflict over community care homes this may well be so.

Planning is an important state regulatory mode through which the production of space in capitalist society is both stabilised and valorised in the interests of accumulation within the land economy (Scott, 1980). The frequent incongruence of urban planning and community care policies in Anglophonic countries therefore has serious implications for the success or otherwise of deinstitutionalisation. A proper integration of the deinstitutionalisation process with the (regulated) social production of space would ensure that all policy regimes which manage the built environment – including planning, health, environmental and building controls – enshrine the principle of locational freedom for community care homes (Gleeson 1996c). This goal of policy integration was briefly pursued in Australia during the early 1990s with promising results. At this time, the national government extended its urban planning programme[3] to include the process of deinstitutionalisation, resulting in a series of well-coordinated, new community care projects in major cities (National Capital Authority, 1996). This highly innovatory attempt at policy coordination foundered, however, in early 1996 when a newly elected neo-liberal government withdrew from national spatial planning altogether.

The observed failure of most Anglophonic states to achieve this policy integration has left planning vulnerable to 'capture' at the local level by interests which are hostile to community care. In this scenario, commonly observed within the USA (e.g., Kindred *et al.*, 1976; Steinman, 1987), Canada (e.g., Dear and Laws, 1986), Britain (e.g., Burnett and Moon, 1983), Australia (e.g., Gleeson, 1996c) and New Zealand (e.g., Shannon and Hovell, 1993), planning becomes a realm of institutional practice where host community hostility is privileged and the right of disabled people to social participation through choice of living environment is curtailed. Even if attempts are made at the supra-local level to remove land use restrictions on care homes, local communities may be able to overcome this protection through resort to other aspects of built environment regulation.

This problem has been clearly exposed recently in the US where local communities have circumvented the Supreme Court ruling on care homes by invoking a range of building controls in order to exclude such facilities. In sum, the problem observed is the failure of states to ensure that their own community care programmes are protected within the public policy realms that condition the production of social space. As the commentators Dear and Wolch (1987) and Joseph and Hall (1985) have observed for the USA and Canada respectively, this failure has allowed wealthy and articulate neighbourhoods to use planning (and other) controls to exclude residential facilities for service dependent people, thus engendering a spatially and socially uneven development of community care networks. I will return to the issue of locational conflict later in this chapter.

Social policy restructuring

The neo-liberal agenda

The incongruence between community care ideals and the reality of urban regulation in advanced capitalist countries is clearly inhibiting the realisation of enabling environments for disabled people. But a further threat to community care has emerged within the social policy sphere itself through an increasingly pervasive and profound restructuring of Western Welfare States, especially those in the Anglophonic world (notably the United States, Britain and New Zealand) where a succession of 'New Right' governments have held power in various times and places since the late 1970s (Johnston, 1993; Thompson, 1990). In broad outline, these changes have been largely (though not wholly) informed by neo-liberal political philosophy, involving, *inter alia*, real reductions in public welfare spending, the shift from universal to targeted forms of public assistance, and the commercialisation of health and welfare services (Barretta-Herman, 1994; Loader and Burrows, 1994). In countries such as Britain and the USA, this process of welfare restructuring has been simultaneous in many regions with the introduction of deinstitutionalisation programmes and the development of community care networks.

Many critics of community care in both Europe and North America have argued that neo-liberal governments have used the policy as a strategy for reducing the costs of welfare provision (Bennie, 1993; Dear and Wolch, 1987; Eyles, 1988; Jary and Jary, 1991; Morris, 1993a). The charge is that deinstitutionalisation has occurred without the simultaneous development of adequate, publicly funded community care networks, and the policy has therefore simply been an excuse for governments to reduce their expensive and increasingly politically sensitive commitments to institutional social support. Thus the continuing enthusiasm of many governments for community care has only raised the suspicions of disability advocates. By deploying the sort of neo-liberal communitarian rhetoric referred to in the previous chapter, critics feel that many national and regional governments have managed to disguise as progressive social policy a more unpalatable reality; notably, a shift in costs of care from the state to individuals, families and local communities.

In the United Kingdom, Clapham *et al.* (1990) have exposed both the inadequacy of care *in* the community (i.e., through funded, professional services) and the parlous state of care *by* the community (i.e., by relations, friends and volunteers). The latter concern refers to the increasing role that communal (i.e., kinship, friendship and neighbourhood) networks have been expected to play within community care in recent decades. Clapham *et al.* (1990) argue that the British government has used deinstitutionalisation to shift the 'burden of care' from public sector agencies to voluntary organisations and informal (i.e., unpaid) carers, usually women.

The failure of many states to fund a full range of independent living options for disabled people has had oppressive consequences for many women. Feminist contributors to disability studies have pointed out that community care too often means family care, and a consequent increase in the domestic burden already shouldered unequally by many women (e.g., Morris, 1993a, 1993b; Parker, 1993). As was seen in the previous chapter, the responsibility for supporting a disabled household member generally falls upon women. Moreover, the duty of care frequently means that women carers are forced to either downgrade their labour force status or withdraw from paid work altogether. In this sense, the reduction of community care to family care sets a double dependency trap for disabled people and women carers. A set of distinct socio-structural forces – disabling labour markets, neo-conservative social policy and partiarchal household relations – combine in such circumstances to reinforce the dependency of women on men, and disabled people on non-disabled relatives and friends. Living within a double-dependency relationship is likely to compromise the life chances of both care giver and disabled recipient. For the latter, it is hardly an enabling environment: 'using friends or relatives as unpaid carers means that the disabled person is unlikely to be able to play an equal role in personal relationships or to participate fully in society' (Morris, 1993a: 27).

In 1988, Eyles observed of Britain that it was 'a misfortune of timing that the desire to develop community care strategies coincided with the fiscal crisis of the British state' (1988: 53). Eyles argued that during the 1980s, the British government failed to fund the full development of community care infrastructure and programmes in the mental health sector, resulting in the reinstitutionalisation of many released patients, and heavy socio-economic and therapeutic strains on other service users. These problems closely reflected the deficiencies in US community care networks which had been pointed to by a number of observers in that country during the 1980s and early 1990s (e.g., Dear and Wolch, 1987; Grob, 1995; Smull, 1990; Wolch and Dear, 1993). More recently, Britain has enacted the NHS and Community Care Act (HMSO 1990), which, among other things, required the production of a community care plan for all areas in England and Wales within two years (Martin and Gaster, 1993).

However, in spite of the exhortatory rhetoric concerning the need for 'individual empowerment' and 'high quality care' of the Act and antecedent reports, a range of observers have pointed to lingering problems in Britain's local community care networks, including under-resourcing, user exclusion, increasing burdens on informal carers, and poor service delivery (e.g., Abberley, 1993; Baldwin, 1993; Bewley and Glendinning, 1994; Ford, 1996; Parker, 1993; Smith *et al.*, 1993). As one disability activist has recently put it,

> the Community Care Act, accompanied as it has been by political rhetoric about independence, choice and control and backed by the belief in the power of market forces to produce it, has failed to break the chains which bind us into unnecessary dependence. It gives disabled people no rights.
>
> (Davis, 1996: 127)

In Australia, staff problems, including poor training, high turnover and low morale, have been cited as serious deficiencies in the country's community care networks (e.g., *Sydney Morning Herald*, 28 January 1997: 1). In New South Wales, the State Government's Community Services Commissioner observed in 1996 that community residences had become a 'favourite source of jobs for backpackers working for a few months to save for the next leg of their journey' (*Sydney Morning Herald*, 30 November 1996: 3).

Wilmot (1997) believes that the British legislation has very little to do with community values, and is, in fact, more concerned with the values of individual choice. In support of this claim, he cites Walker's (1989) view that the former Conservative Government's agenda for community care emphasised 'increased emphasis on self-help and family support, extension of the market and commodification of social relations' (1997: 31). As Wilmot points out, 'The first of these does not necessarily relate to community and can function as an alternative to community care', while 'The latter two are antithetical to community' (ibid.). Moreover, Bewley and Glendinning (1994) report the increasing anger and cynicism among disabled people in Britain who have been involved in local community care planning exercises. According to these observers, many users experience community care policy settings as *disabling* rather than *enabling* political environments (see George, 1995).

In the era of neo-liberalism, the goal of 'improved resource utilisation' is now an almost mandatory shibboleth in the social policy discourses of many advanced capitalist nations. However, in a range of countries, including Britain (e.g., Eyles, 1988; Jary and Jary, 1991), New Zealand (e.g., Kearns *et al.*, 1991, 1992; Kelsey, 1995) and the United States (e.g., Dear and Wolch, 1987), it has been argued that the prioritisation of cost savings over service quality and extent by public agencies has been a major reason why community care networks have never materialised on a scale sufficient to support the needs of many socially dependent persons. In Britain, Lewis and Glennerster (1996) have argued that the main purpose of new community care strategies in the 1990s was to rein in social security spending. The under-resourcing of deinstitutionalisation and replacement support networks means that community care is hardly likely to provide the sort of enabling environment which disabled people require in order to secure their needs for material well-being and social participation.

For example, a recent study of British community care housing by the Joseph Rowntree Foundation painted a bleak picture of the accommodation choices open to disabled people. The Rowntree Foundation report found that much of the social rental housing used for community care had been relegated 'into a stigmatised and residual sector catering for those who have no other choices' (*Guardian*, 2 July 1997: Society 9). The evidence was that disabled people were frequently shifted from institutions into accommodation that was characterised as 'grotty flats on high crime estates' (ibid.) – hardly the enabling residential settings that disability activists and advocates have struggled for. Again, government fiscal stringency was identified as a major cause of the housing problems. As the Rowntree Foundation put it:

Community care policy makes many claims about enabling people to live more independently and direct the course of their own lives. These claims do not square with the findings from the ... Foundation's Housing and Community Care Programme. There is much poor quality accommodation, haphazard funding of support services, lack of access to housing to those who want to move from a family or institutional setting, and reliance on a limited range of specialist service provision.

(Joseph Rowntree Foundation, 1997: 1)

In Canada, Cormode (1997) reports that Ontario's neo-liberal provincial government has recently introduced new eligibility criteria that restrict disabled people's use of accessible transit services. Cost cutting appears to be the main rationale for the change: 'The official reason given for the new service criteria was limits on funding from the provincial government' (Cormode, 1997: 389).

Allied to the threat of under-resourcing is the set of potential problems that human service commercialisation raises for community care. In many advanced capitalist nations community care facilities for disabled people are increasingly provided by voluntary agencies and for-profit organisations (Jary and Jary, 1991), reflecting a growing preference among many governments for the contracting out of human services (for the USA, see DeHoog (1984) and the collection edited by Demone and Gibelman (1989); for Britain, see Lewis and Glennerster (1996) and the work edited by Allen (1992)). In many national contexts – especially within the USA (McGovern, 1989), the UK (Allen, 1992; Leat, 1995; Malin, 1987; Smith *et al.*, 1993) and New Zealand (Abbott and Kemp, 1993; Le Heron and Pawson, 1996) – the transition to community care has coincided with a shift to contracted arrangements and the replacement of public service providers by voluntary and for-profit agencies.

The transition from centralised, publicly provided care has been encouraged by a diverse set of socio-political interests, ranging from disability advocates and social policy commentators (e.g., McGovern, 1989; Smith *et al.*, 1993) who have emphasised the empowerment benefits for users of decentralised, 'communal' (i.e., local, non-state) care to neo-liberal theorists (e.g., Savas, 1982; Foldvary, 1994) who have tended to stress the efficiency gains, the improved fiscal accountability and the enhanced consumer choice that supposedly flow from contracting out. A key organisational feature of the contracting system is the functional and fiscal split between public purchasers of human services and the private (i.e., voluntary and for-profit) providers of these commodities (Allen, 1992). This 'funder-provider' model of human service contracting has been a keystone of national health restructuring in both Britain (Baldwin, 1993; Browning, 1992) and New Zealand (Blank, 1994, Fougere, 1994) during the past decade.

Wolch's assessments (1989, 1990) of the shift to 'voluntarism' by governments at all levels in the USA during the 1980s point to an organisational convergence between voluntary and for-profit providers. According to Wolch, increased competition for service contracts between for-profit and not-for-profit agencies forces many voluntary and charitable bodies eventually both to adopt an entrepreneurial ethos and to emulate market organisational structures. This 'marketisation' of the not-for-profit sector is reflected in the tendency of voluntary groups to (re)prioritise economic efficiency over other organisational goals (a process that Wolch (1989: 216), after Kramer (1986), refers to as 'goal deflection'), leading, *inter alia*, to the adoption of market pricing for their services (i.e., user fees) (see also Smith and Lipsky, 1993).

The privatisation of care: 'a deadly gamble'?

If the contracting system encourages a concern for profit – or at least, a greater emphasis on fiscal efficiency – among voluntary sector providers of human services, what might be the implications of this shift for community care? One cautionary tale is surely provided by the saga of private nursing home care in the United States which has been carefully documented by Vladeck (1980). By 1980, the commercialised US nursing home industry – ostensibly a 'community care' network for the frail elderly with more than 17,000 facilities – had deteriorated to the point where it represented a grave threat to the health and well-being of its users. Vladeck's account catalogues the horrors that arose from the systemic 'indifference, neglect and physical abuse of patients' by home operators (1980: 4). He blamed this deterioration on a combination of poor monitoring of the sector by governments (some two-thirds of the industry's revenue was publicly sourced) and the 'financial chicanery' of private facility operators, many of whom it seemed were willing to inflict gross indignities on home residents in order to maximise profits (ibid.). As Vladeck concludes, 'The experience of [US] nursing home policy teaches many useful lessons that can be applied to other areas of public concern' (1980: 5).

By 1997, there was evidence to show that the situation in many nursing homes in the United States had not improved much since Vladeck's study in 1980. One prominent disability activist related the following disturbing view of contemporary nursing home practices:

> I spoke with two women who had worked in medicaid funded nursing homes where they say the residents/inmates were literally starving. One of those women was fired for feeding the hungrey (sic) with food she had purchased. They said people were rarely bathed and lived with horrid bedsores on their bodies. No one is allowed to have sex (except those raped by employees) and you can't even have a beer on your birthday.[4]

During the 1990s, ADAPT activists have campaigned against the tendency of State and Federal governments to use nursing homes as proxy institutions for disabled people (Dorn, 1994). ADAPT has argued that a significant percentage of Medicaid funding should be diverted from nursing homes into attendant care programmes that would help disabled people to live independently. In November 1997, the United States House of Representatives debated a bill (H.R. 2020) that would establish a national personal assistance policy, and thereby greatly improve the residential choices available to disabled people. At one ADAPT demonstration in support of the bill, 450 protesters carried placards proclaiming 'I'd rather go to jail than die in a nursing home'.[5]

Moreover, as early as 1988 evidence emerged in California to show that the shortcomings of privately provided nursing home care were being reproduced in new community-care facilities designed specifically for disabled people. An investigation then by the *Los Angeles Times* into community care for intellectually disabled people revealed 'a widespread pattern of lethal neglect, physical and sexual abuse and financial exploitation of retarded people living in privately-run facilities throughout the state' (*Los Angeles Times*, 8 January 1989: 1).

On the basis of much extremely disturbing evidence, the newspaper concluded that 'placing patients in privately run facilities can be … a deadly gamble' (ibid.). Parents and guardians of disabled people were doubtless quite aware of these depredations and the newspaper reported that many reacted by 'fighting to keep their children in state hospitals and out of privately-run homes' (ibid.). Throughout the Western world, the failure of many governments to resource and manage community care programmes adequately has diminished popular support for deinstitutionalisation and has even encouraged some disability advocacy groups to call for the establishment of new and enhanced institutional facilities (Gleeson, 1996c).

Finally, a glimpse of the problems that may arise from for-profit provision of disability services was recently provided in Australia. In January 1997 it was alleged that a privately run institution for intellectually disabled children posed 'serious risks' to its 60 residents and could not guarantee their safety (*Sydney Morning Herald*, 28 January 1997: 1).[6] A report by the New South Wales Community Service Commission catalogued a series of human rights infringements at the facility, mostly related to its 'extremely poor' management. The report stated:

> It is a geographically isolated service providing seriously sub standard care that, instead of ensuring the well-being, development, care and safety of its residents, exposes them to danger and systematically fails to provide for their needs. Many of the regular occurrences at the service would horrify outsiders, yet residents and staff are expected to deal with them as part of their daily routine.
>
> (Community Services Commission, 1997: 1)

Critics called the institution 'Hell for Children', and the New South Wales Council for Intellectual Disability demanded its closure. The Community Service Commission criticised the fact that one individual was both landlord and administrator of the facility, an apparent conflict of interest that raised disturbing implications about service management and quality.

A cautionary tale: New Zealand's neo-liberal revolution

To date, however, there is little evidence that this lesson has been absorbed, or even noted, in the countries where the contracting of disability services is flourishing. In New Zealand, a country that has been subjected to a radical programme of neo-liberal restructuring since 1984 (Kelsey, 1995), recent and continuing public policy changes have sought to restructure service provision in the health and welfare sectors (Blank, 1994; Boston, 1992; Shannon, 1991). Critical among these changes in health and welfare policy domains has been the imposition of a purchaser–provider split and the increased use of service contracting (Fougere, 1994; Le Heron and Pawson, 1996). On the basis of observed United States experience, Fougere (1994) fears that health costs will rise, a consequence of particular significance to the many fixed and low income users of community care services.[7]

It is also likely that 'for-profit' agencies will play an increasing part in providing community care facilities in the near future. Gleeson *et al.* (1995) have argued that this development may change the nature of community care facilities in New Zealand, which to date have mostly been of modest size (comparable to most residential uses), as private providers attempt to realise economies of scale in service delivery and maximise profits through the provision of larger 'homes'. As Fougere notes of the new contractual model in New Zealand health services, 'Competition may encourage providers to skimp on those aspects of quality least visible to purchasers' (1994: 157). Indeed, in the context of community care, such 'quality supervision' may be a difficult task, given the frequent need for users to be represented by advocates and/or guardians.

Bennie, reflecting upon community care policy experience in Australia and the USA, warns that

> The practice of allowing 'private operators' to set up as residential providers relying solely on residents' benefits has reproduced many of the problems associated with economies of scale that are a feature of hospitals. Because these 'entrepreneurs' do not receive grants by way of contracts, there is often little external monitoring or evaluation. Many operate as large boarding houses with only minimal comforts and an absence of structures support services. Exploitation of residents in these circumstances has been widely reported.
>
> (1993: 20)

In Britain, Finkelstein and Stuart have observed that 'Closing down ... institutions has only meant that the institutional approach to care has moved into the community under the "community care" fiat' (1996: 181). Baldwin (1993) also makes detailed criticisms of community care service standards in Britain. Rea (1995), on the other hand, undertakes a broader appraisal of welfare voluntarism in Britain, and argues that this approach encourages 'unhealthy pluralism' in community care provision. For this author, competition is an 'unhealthy' principle because it undermines the long-term inter-organisational collaboration that is a necessary condition for effective human service delivery (see also Hoyes and Means, 1993).

New Zealand's neo-liberal restructuring process has also impacted upon the country's urban and regional planning system, which was changed profoundly in the early 1990s by the enactment of new legislation – the Resource Management Act 1991 – that, *inter alia*, liberalises land use regulation in a variety of ways (Memon and Gleeson, 1995). On the face of it, this shift to a flexible system of land use control, based on 'performance zoning', seems to realise Young's (1990) ideal of a non-discriminatory planning regime. However, on closer analysis, it is clear that the changes to New Zealand's planning reflect the desire of neo-liberals for a deregulated land use system, rather than any concern to minimise socio-spatial segregation (Gleeson and Memon, 1994, 1997). That is to say, the main aim of the changes is to enhance the profitability of capital by increasing the locational prerogatives of developers while also reducing the transaction costs that arise from public regulation. Again the divergence of social policy ideals from planning policies and practices is instructive.

In a study of the new planning regime in which I participated during 1994, it emerged that flexible land use controls were likely to affect profoundly the provision of community care services (Gleeson and Memon, 1997; Gleeson, *et al.*, 1995). Several of the new local zoning schemes produced under the legislation entirely deregulated community care homes, leaving it open for service providers to decide the extent and nature of such facilities. In many instances this shift will benefit disabled people by freeing up the provision of quality community care services. I refer here to the residential services provided by public and voluntary agencies in small-scale settings that stress the dignity and autonomy of their disabled residents. However, it was clear that the same changes would also benefit the new private sector providers that were being attracted into the community care sector by the neo-liberal restructuring of New Zealand's social and welfare policy regimes. Deregulation would liberate private operators from any public planning controls on the nature and quality of community care facilities, thus further encouraging the development of large private-sector community care facilities.

If large, private facilities eventuate, it is likely that many urban communities will resist their establishment on the ground that they are 'mini-institutions' and therefore do not qualify as legitimate residential land uses. However, the new, liberalised planning controls reduce the extent of public notification for most developments, including care homes, and it is likely that local communities may only be aware of such large 'mini-institutions' some time after their establishment. The potential for host community resentment and *ex-post* locational conflict may be high in such a scenario.

It is obvious that the emerging planning context for community care in New Zealand fosters an economic diversity that differs, perhaps markedly, from the desegregated community life which is central to Young's vision for an inclusive city. While planning reforms in New Zealand may have reduced the ability of local communities to frustrate deinstitutionalisation through NIMBY actions, they have also significantly reduced the capacity of local and regional states to control the nature and quality of community care services. By contrast, a progressive planning policy that sought to encourage enabling residential services would guarantee locational freedom only to those facilities that met carefully defined standards governing the nature and quality of community care homes. While planning controls in the past may have had little to do with the quality of community care services, the complete deregulation of this land use category forecloses on a better informed, and more enabling, approach to development control that would prevent community care being used as a guise for the reinstitutionalisation of disabled people, this time in a new privatised institutional landscape.

Iveson (1998) upbraids Young for encouraging planning systems to adopt a rather undiscriminating vision of diversity that seems to avoid the question of inequalities of economic power. His criticism seems borne out in the New Zealand case where the conjunction of simultaneous neo-liberal reforms to planning and community care has produced a regulatory scenario that prioritises economic diversity over social inclusion.

In summary, the shift to private provision of community care, which is at present occurring in a range of Western countries, may worsen accommodation options for dependent disabled people by encouraging the creation of large-scale 'care' facilities in residential areas that will doubtless be objects for concern and hostility in many local communities. This social conflict further threatens the viability of community care and raises the possibility of (increased) social stigmatisation of disabled people. Moreover, these regressive shifts may be reinforced by any broad-scale deregulation of land use controls in urban and regional planning.

Conclusion: the limits to justice

This chapter examined one area of state policy practice – community care – from the perspective of enabling justice. It was recognised that community care alone cannot deliver justice for disabled people. First, this policy sphere is only of relevance to *service dependent* disabled people, thereby not directly touching the lives of many who sustain themselves by other means (e.g., work and kin networks). Second, residential social programmes cannot address certain critical injustices which disabled people face, including exclusion from labour markets and mainstream political fora. None the less, it is recognised that community care has the potential to diminish the specific injustices which many disabled people have faced in the past through incarceration within the institutional spaces of social dependency.

The analysis then reviewed a set of further threats to the policy of community care in a range of advanced capitalist nations. These problems included the frequent opposition of local communities to residential social programmes (the NIMBY syndrome), the friction between planning regulation and community care policy practice, and the broader restructuring of state health and welfare activities by neo-liberal governments.

Community care holds the promise of moderating injustice for many disabled people by offering service dependent groups an expanded choice of valued living settings. However, the realisation of this promise is dependent upon a state policy practice which will produce a new and enabling social space of care in capitalist societies. As pointed out in the foregoing analysis, there are considerable structural impediments to the production of this new care landscape. In particular, the prevalence of NIMBY sentiments, and the failure of states to integrate social policy and urban planning regulation, may mean that the new landscape of care is as concentrated and socially isolating for its 'inhabitants' as that which it replaced (cf. Kearns, 1990; Laws and Dear, 1988). Indeed, the recent North American experience suggests that deinstitutionalisation often simply means *trans*-institutionalisation as service dependent people migrate from traditional institutions to gaols, mainstream hospitals and/or the clusters of over-stretched facilities concentrated within inner city 'service ghettos' (Bennie, 1993; Dear and Wolch, 1987; Grob, 1995; Wolch and Dear, 1993).

Moreover, the socio-political obstacles that have been raised in the path of community care have discouraged some governments from undertaking full-scale deinstitutionalisation. This situation is all the more lamentable for the fact that many of these obstacles, as I have shown in this chapter, are created by states themselves. In the State of Victoria, Australia, a series of policy setbacks, at least some of which are attributable to government fiscal conservatism, seem to have slowed the process of deinstitutionalisation. As evidence of this, the State Government decided recently to establish a new institution, the first to be built in Victoria for twenty years. Ostensibly, the move is part of a positive development that will close one ageing and decrepit institution which currently houses 350 people. However, not all residents will be deinstitutionalised, and many will be placed in a new, 105 bed 'congregate care' facility that is to be established on the site of the old institution. Local disability groups were outraged by the decision, and in late 1997 lodged a complaint against the State under the Victorian Equal Opportunity Act (Ripper, 1997).

In countries that have restructured health and welfare services along neo-liberal lines, the prospect of commercial service providers establishing large-scale facilities in residential areas means that the transition to community care may simply involve the *re*production of institutional space (Bennie, 1993; Elliget, 1988). In this case, community care represents an 'enabling environment' only for the neo-liberal interests which demand reductions in public spending through human service restructuring that seeks little more than 'improved resource utilisation'. (This aspiration is at least partly revealed in recent economic analyses that make the theoretical case for the private provision of social services (e.g., Foldvary, 1994), stressing efficiency concerns over distributional or cultural aims.)

The foregoing discussion has identified several powerful political-economic constraints on community care, a policy practice which aims to reduce, or even overcome, the limitations of welfarist justice for disabled people. As I have shown, there is the danger that such structural impediments will condition the production of a dependency landscape that entrenches, or even worsens, the disadvantage of some disabled people. This is not, however, to declare the futility of community care as a strategy for lessening the socio-spatial injustices experienced by disabled people. Rather, the foregoing discussion exposed the emancipatory limits of community care *as it is at present practised* by most advanced capitalist states. From this it is clear that the real emancipatory limits to community care will not be reached (and therefore known) until this form of policy practice engages the deeper processes that condition the production of social space. This chapter has implicated at least one such point of engagement between community care and structural processes by highlighting the need for greater integration between community care and urban land use regulation. The threat raised by neo-liberal social policy requires a broader political engagement which seeks to wrest control of state ideology from econocrats and thereby restore human values to public policy.

9 The regulation of urban accessibility

Introduction

This chapter will explore the urban geography of disablement through a case study of accessibility regulation. As with the previous case studies, this empirically based investigation will use historical-geographical analysis to explain how specific dimensions of disablement arise, and are reproduced, through the interplay of structural, institutional and contextual conditions. In this case, the spheres of accessibility regulation – access laws, building standards, rights-based guarantees – are examined critically, with a view to explaining the origins of the inaccessibility of capitalist cities. I will also show how this form of oppression is both reproduced and challenged through institutional and political practices.

Most Western nations have now enacted laws and codes that aim to improve the physical accessibility of cities for all users (Imrie, 1996b). Such regulations recognise the specific (and long ignored) mobility needs of disabled people (Napolitano, 1996). While a considerable evaluation literature has emerged within Western nations to assess aspects of these legislative initiatives, such as policy coherence and the adequacy and consistency of design standards, there has been little social theoretical analysis of how these regulatory regimes have fared in practice. In particular, the issues of regulatory compliance, and the political economic context for this, have rarely been addressed.

Imrie's (1996a) important recent monograph extends this analysis to a broader international context through a comparative examination of access regulation in Britain and North America. While the breadth of Imrie's comparative frame is considerable, the geographical exclusions in his analysis are none the less considerable (and understandable). As Imrie himself acknowledges, social scientific understanding of state disability policy practice is mostly limited to 'the USA, the UK, Sweden, and one or two other western European countries' (1996a: 176).

To date, for example, there has been no critical theoretical analysis of access regulation in Australia or New Zealand, even though both countries have at times taken internationally significant initiatives in aspects of disability policy (Gleeson, 1998). In New Zealand, access problems have been raised by a range of advocacy groups and policy commentators (e.g., Cahill, 1991; Wrightson,

1989) in recent years, but there has so far been no critical analysis undertaken of that country's comprehensive accessibility regulations.

As related in the previous chapter, New Zealand in the past decade experienced a radical transformation of its public policy regimes along neo-liberal lines. New access regulations have been enacted within a policy environment increasingly defined by emphases on deregulation, administrative flexibility and public sector cutbacks. The profundity of these neo-liberal reforms and the singularity of the public policy environment they have created mean that understandings of accessibility regulation derived from studies of overseas contexts cannot readily be applied to the New Zealand case. However, given the increasing influence of neo-liberal prescriptions for public sector reforms in many Western countries, it would seem that New Zealand provides a valuable and instructive case setting for analysis of regulatory change (Kelsey, 1995). Critical examination of how accessibility regulation has fared in New Zealand may provide conclusions which are relevant to other national contexts that have not experienced such thorough-going structural changes.

Recognising these inadequacies in the literature, my aim in this chapter is to explore the socio-political context for access regulation in capitalist cities. Specifically, the chapter will identify both the key theoretical questions that might inform further studies of access law compliance and some of the compliance problems that occur in capitalist cities through a case study of regulation in one urban area, Dunedin, New Zealand. Data for the case study were derived from a set of primary and secondary sources obtained during 1995. The principal primary information source was a set of interviews with twenty people with a first-hand knowledge of access regulation in New Zealand, including several disabled persons.[1] This exploratory analysis is by no means exhaustive – further empirical research is needed in order to better understand the forces that shape accessibility regulation in specific capitalist cities.

The chapter is structured as follows. The first section outlines the theoretical context for the study by briefly reviewing the small amount of critical literature on accessibility. Next the discussion outlines the New Zealand legislative and policy contexts for accessibility regulation. Following this, a brief summary of the Dunedin case study is presented. A final section then considers the theoretical and policy implications of the research.

The production of disabling space

Policy context

Although most Western countries now have in place some form of building and planning legislation which attempts to counter the problem of inaccessibility, there is accumulating evidence to show that such policies are generally failing to reduce or prevent discriminatory urban design (e.g., Bennett, 1990; Gilderbloom and Rosentraub, 1990; Imrie, 1996a, 1996b; Vujakovic and Matthews, 1992, 1994). Access legislation is often opposed by development capital, and governments tend to be less than rigorous in its enforcement. Recently in the United States, powerful corporate interests have argued before the federal judiciary that the Americans with Disabilities Act (ADA), by requiring businesses to provide wheelchair access, is an unnecessary restriction upon private property rights, and therefore an infringement of the Fifth Amendment to the US constitution (Helvarg, 1995).

In Britain, Imrie and Wells (1993a, 1993b), have shown how the Thatcher government during the 1980s progressively relaxed central controls on accessibility standards, and encouraged a mood of regulatory voluntarism among local authorities (which bear the primary responsibility for enforcing access codes). The authors argue that many local authorities subsequently gave little policy priority and few resources to accessibility responsibilities. The national lethargy on access policy was attributed in part to the flourishing climate of local growth politics, and the consequent anxiety of individual councils that superfluous building regulations would frighten away increasingly mobile development capital (Imrie and Wells, 1993a, 1993b). Progressive fiscal cutbacks by central government had reduced the overall regulatory capacity of local states.

In a more recent analysis, Imrie (1996a) shows that while awareness of access issues in Britain has increased at the local authority level in the past decade, the ability of councils to regulate built environment change for social ends has been greatly undermined by the central government's deregulation of planning and building control. The Department of the Environment has encouraged a climate of regulatory voluntarism through a series of guidance circulars and rulings that have discouraged councils from using access as a dimension of development control. Moreover, in a major concession to commercial lobbies, the British government in early 1995 sought to cushion the impact of new access regulations for businesses by limiting the amounts of both money and time that must be expended in compliance. Disability advocates reacted to the new bill with outrage. One activist described it as 'a set of half measures which were neither comprehensive nor legally enforceable' and argued that 'many employers will remain free to exclude and discriminate against disabled people' (*Guardian*, 13 January 1995: 7).

Reflecting also on the British experience, Thomas (1992) has invoked Illich's notion of 'disabling professionals' to describe the frequent tendency of public planners to undermine access laws through policy (non)practice. According to Thomas, planners are often both ill-informed about access laws and lethargic, or even obstructionist, in their enforcement. Recalling the analysis of the previous chapter, this reactionary tendency is doubtless another instance of the problems caused by the separation of urban planning and social policy in state practice.

Theorising accessibility

Accessibility has been poorly theorised in the spatial sciences. Within Geography, one struggles to find any recognition, let alone analysis, of access issues for disabled people. (Important work has been done, however, by geographers on access issues facing other social groups, notably women (e.g., Fincher, 1991; Rose, 1989).) Outside Geography, there have been several examinations of the inaccessibility problem which confronts disabled people in contemporary cities – for example, in Urban Planning (e.g., Bennett, 1990) and in Architecture (e.g., Lebovich, 1993; Leccese, 1993; Lifchez and Winslow, 1979; Kridler and Stewart 1992a, 1992b, 1992c). The attention given to access issues in Architecture in recent decades has been encouraged by new advocacy lobbies and professional fora that have sought to emphasise 'barrier free' design (e.g., Wrightson, 1989). These investigations have certainly influenced public urban policies on disability issues in the form of access legislation and inclusive building codes. However, a major weakness of the access literature is its frequent tendency to reduce the social oppression of disablement to a physical design problem.

Indeed, the general treatment of the spatial question within disability studies has tended to display a crude materialism in which the arrangement of the built environment is seen as the principal source of disablement. In this view space is reduced to an inanimate configuration of material objects, and its sense of sociality is lost to analysis. The implied phenomenal form of space, the city, simply becomes a static diorama, freed from the social structures which created it, and the issue of disablement is reduced to a dilemma of access.

Urban geographic analysis could contribute to an enlarged debate on the origins of inaccessibility. Indeed, it does seem that this enabling potential is at last being drawn upon through new studies of access issues (e.g., Gant, 1992; Gant and Smith, 1990; Golledge, 1993; Gleeson, 1997; Imrie, 1996a, 1996b; Imrie and Wells, 1993a, 1993b; Olson and Brewer, 1997; Vujakovic and Mathews, 1994).[2] These analyses have all explored different aspects of the production of inaccessible space, ranging from the mobility experiences and strategies of disabled people (Gant, 1992; Gant and Smith, 1990; Vujakovic and Mathews, 1994; Matthews and Vujakovic, 1995), to the role played by regulatory and political economic forces in determining levels of urban accessibility (Gleeson, 1997; Imrie, 1996a, 1996b; Imrie and Wells, 1993a, 1993b). Much more can be done, however – an issue that I will return to in more detail in the next chapter.

The recent influential work of Golledge (1990, 1991, 1993, 1996) also critically examines the inaccessibility of the contemporary Western city for disabled people. However, he departs from a historical-geographical approach in important ways. Golledge argues that disabled people inhabit 'distorted spaces' (e.g., 1993: 64): in fact, he envisages a unique 'world of disability' (e.g., 1993: 65) that corresponds to the constricted time-space prism of the disabled individual. In another recent geography of disability, Vujakovic and Matthews (1994: 361) echo this socio-spatial ontology with their stress on the 'contorted, folded and torn' environmental knowledges of disabled people. This approach contrasts with the social model because these so-called 'worlds of disability' are seen to have a primarily pathological genesis, located in the deficiencies of the disabled body rather than in social phenomena. These deficiencies are exaggerated, but not caused, by the social arrangement of space. Environmental modifications which seek to increase access for disabled people are explained as 'efforts to *compensate* for disability' (Golledge, 1993: 64, emphasis added). The clear implication is that disability is a set of physiologically given deficiencies, rather than socially created limitations, for which society seeks to compensate the individual through environmental design concessions. From a historical-geographical perspective this ontology is fundamentally flawed and overlooks the socio-spatial production of disability (cf. Dorn, 1994; Wolpert, 1980).

Recently, the geographic theorisation of inaccessibility was greatly advanced by the publication of Imrie's (1996a), *Disability and the City*, the first major geographic analysis of disability in published form. The book attempts an international analysis of how disabling (inaccessible) cities are produced, though the empirical materials are mostly drawn from Britain and, to a lesser extent, the United States. This analysis centres upon the role played by key agents and institutions – notably the state (central and local) and building professions (planners, architects, builders) – in the creation of built environments that ignore the fundamental needs of disabled people. Much of this discussion is sourced in the author's own considerable primary investigations of policy practice by local authorities in Britain.

Imrie begins his policy-based analyses of urban inaccessibility with a broad theoretical understanding of disability as a form of social oppression. In this, he draws upon the growing, critical sociology of disability in order to link inaccessibility to the other dimensions of a larger experience of social oppression shared by most disabled people (e.g., poverty and political marginalisation). This broad theoretical frame includes political economic themes – centring on deregulation and the ascendancy of local growth politics – which Imrie sees as critical structural constraints on public policies for disabled people. Imrie then considers the regulation of the built environment, focusing both on the origins of inaccessibility and the legislative and institutional initiatives which have sought to combat disabling urban design. As seen earlier, Imrie shows that access policies have thus far carried little regulatory weight in Britain.

Finally, Imrie considers the political project of reconditioning the production of built environments in order to improve accessibility. As he notes, the attempts of local authorities in Britain to consult with disabled people in order to develop more inclusive regulatory policies have thus far not been encouraging. While the limitations of the 'advisory approach' are manifold, strategies for a more radical, transformative politics by disabled people are far from obvious in an era of pervasive neo-liberalism.

Policy implications

What are the policy implications of these theoretical critiques of access laws in market societies? First, it seems clear from observed practice that several institutional and socio-political forces seem to undermine the effectiveness of access legislation. These forces include commodity relations which devalorise disabled people as workers and consumers; discriminatory cultural practices that reinforce disabling constructions of public space; the fiscal problems of local authorities;

and the fact that the capitalist land economy emphasises profit over other potential social objectives, such as inclusive design.

Moreover, even if access laws were to be fully applied, their ability to reduce the social oppression experienced by disabled people has serious structural limits which arise from the organisation of capitalist societies. While access laws have a great value in reducing the mobility friction in everyday life for many physically impaired people, they do not address the deeper socio-spatial dynamics – such as the commodity labour market or the land economy – that produce disabling environments. Better building standards and new modes of mobility, for example, will not *on their own* revalue the labour power of all physically impaired people. Such strategies can reduce the friction of everyday life for disabled people, and must be defended for this, but they will not solve the dynamic socio-spatial oppression of disablement.

Applying Fraser's (1997a) analysis, access regulation reflects the weakness of any 'surface reallocation' of resources towards disadvantaged social groups. (Fraser particularly has in mind affirmative action policies for women, but her observation is equally valid for access laws.) As she points out, any such strategy is at best ameliorative and cannot engage the underlying patterns of social organisation that produce injustice: 'Leaving intact the deep structures that generate … disadvantage it must make surface reallocations again and again' (1997a: 29). According to Fraser, therefore, we should not be surprised if access improvements to built environments are constantly undermined or even eliminated over time.

This theme of 'structural-institutional limitation' on state regulation forms the theoretical background for the following case study of access controls in New Zealand. However, cultural and material structures by definition cannot be described and measured in the way that material objects can, as their presence in day-to-day affairs is felt through concrete representations (e.g., 'markets' may be experienced as money and individual acts of exchange) (Sayer, 1992). Thus, as will be evident in the following empirical analysis, the influence of political-economic factors on disability regulation in New Zealand is implied rather than confirmed – the implication might well be strengthened if further studies could develop the empirical 'picture' drawn in this case study.

Access regulation in New Zealand

The world's best accessibility legislation?

Over the past twenty years, the New Zealand government has promulgated a series of acts and regulations that have aimed to improve the accessibility of built environments for physically disabled people. These legislative and administrative initiatives have gradually increased the strength of access legislation, leading to claims at a 1993 national conference of architects that New Zealand had 'the best legislation for accessibility in the world' (*Dunedin Star Midweek*, 10 November 1993: 1). The accumulation of laws and codes concerning different aspects of access has certainly created a complex and labyrinthine area of state regulation. It would be pointless, therefore, to attempt an exhaustive explanation of this regulatory field in the present discussion. The bulk of the access regulation is contained in two statutes: the Building Act 1991 (BA) and the Human Rights Act 1993 (HRA). For the purposes of the present analysis, a summary outline of these two legislative frameworks will suffice.

The Building Act 1991

The newly amended BA (s.47A) establishes that *building accessibility* is to be measured in three ways. First, a building must provide 'reasonable and adequate provision by way of access'. Next, this access must allow a disabled person 'to visit or work' in that building. Finally, accessibility means that a disabled person must be able to 'carry out normal activities and processes in that building'. The tests are integrative – a building must meet *all three* provisions if it is to be judged accessible under the BA. The regulations apply to all new buildings and associated spaces (e.g., driveways, connecting passages, etc.), as well as to structures which undergo substantial alteration. The access standards under the BA apply only to the public sphere (and not all of this) – privately owned domestic dwellings, small industrial buildings, various agricultural buildings and other minor structures are exempted under the legislation. It is important, therefore, to note that this sphere of regulation probably affects only a minor portion of the built environment of New Zealand's cities.

The BA establishes an elaborate system for regulating the accessibility of public buildings. First, the responsibility for supervising the application of the Building Code, including its accessibility provisions, is vested in the Building Industry Authority (BIA), a central state agency with quasi-judicial power both to interpret control documents and to resolve differences between owners and territorial authorities which arise in the application of controls.

The BA is administered by territorial authorities (TAs) (district and city councils). For TAs, the legislation takes the practical form of a Building Code, a set of construction design standards which cover, *inter alia*, stability, fire safety, access, and energy efficiency. The Code is based on flexible 'performance controls' and does not dictate design standards in any detailed way. Builders must meet the design outcomes specified in the Code. Territorial authorities issue consents under the Building Code for the construction and occupancy of buildings. Given the Code's performance-based approach, the onus for TAs is on ensuring that building owners provide evidence of compliance, rather than on conducting these activities themselves. Building Officers employed by the TAs have the major responsibility for ensuring that building owners comply with the Code. These Officers may undertake site inspections during construction or alteration works in order to monitor compliance.

A major area of ambiguity occurs in relation to Code compliance for alterations to the use or structure of existing buildings. First, owners are obliged to notify the relevant TA if the use of a building changes. The TA must be satisfied that, in its new use, the building still complies with access provisions. However, the wording in the BA allows for flexibility in compliance by stating that, after a change of use, a structure must comply with the Code 'as nearly as is *reasonably practicable*' (s46(2)(a)). The same test applies to alterations to existing buildings.

Generally, the TAs have the power to decide a waiver or modification of the building code. However, waivers for access controls may be granted only by the BIA. Importantly, this limitation applies to new buildings only and there is some leeway for TAs to grant waivers for access matters in existing buildings. The legislation provides several means through which TAs may penalise breaches of the Building Code.

The Human Rights Act 1993

The Human Rights Act 1993 (HRA) provides an additional layer of accessibility regulation. This statute consolidates and extends existing human rights law in New Zealand.[3] It is intended that the HRA will eventually override all other legislation, including the Building Act, although this will be delayed until 1999 and the completion of a systematic review of all laws to comply with human rights principles. Enforcement of the HRA is the responsibility of the Human Rights Commission.

The HRA does not provide blanket protection against all forms of social discrimination. As Stewart notes, 'The bill does not outlaw all discrimination – it prohibits discrimination on certain grounds (including disability) in certain specified areas' (1993: 8). The discrimination grounds include sex, marital status, race, colour, age, religious or political belief, and disability. For disabled people, this prohibition applies to the following social areas:

- employment;
- access to public places;
- education;
- provision of goods and services; and
- housing and accommodation.

Crucially, the HRA may (this is yet to be demonstrated) extend accessibility legislation to private sphere domains where housing and other services are provided. Under the HRA, discrimination essentially means unequal, less favourable treatment[4] for a person, and it could therefore be argued that the inaccessible design of housing (especially commercial accommodation) is one way in which people can be treated unfairly. Of central importance to the present analysis are the sections which guarantee the access of disabled people to places, facilities and vehicles. It is unlawful to refuse to allow anyone access to or use of places or vehicles which the public is entitled to use.

However, as with the building legislation, the HRA is rather equivocal on the subject of compliance – indeed the two statutes share the same ambiguous verbiage in this respect. In addition to the prohibited grounds of discrimination, a parallel set of 'exceptions' is provided. For example, the HRA forbids discrimination in access to public places and vehicles, except where 'it would *not be reasonable* to require the provision of ... special services or facilities' in order to secure accessibility (emphasis added). Similar 'reasonable accommodation' exceptions apply to other potential areas of discrimination, such as employment and housing. Clearly, this potentially ambiguous aspect of the HRA, which is yet to be clarified in case law, has the potential to complicate, and thereby undermine, the enforcement of, and compliance with, the legislation (Stewart, 1993).[5]

Given the newness of the HRA, its potential to reduce the discriminatory design of New Zealand's cities is yet to be revealed (some initiatives taken by disabled people under the HRA are explored later in the discussion). In contemporary New Zealand, the Building Act and its associated Code remain the principal legal and administrative means for ensuring the accessibility of public buildings and spaces.

New Zealand's access framework in international context

In many respects the dualistic access framework in New Zealand is unique for its combination of a US-style rights-based approach (the HRA) and a more traditional British regulatory approach, resting on building controls administered by local councils. The HRA emphasis on disability reflects the philosophical approach taken in the United States by the Americans with Disability Act, both by emphasising inaccessibility as an infringement of basic human rights – i.e., rather than simply as a 'design fault' that hinders mobility – and by linking the access issue to related problems in areas of housing, employment and social services (Imrie, 1996a). The US legislation then goes on – as the HRA does not – to link very specifically the guarantee of individual access rights to local planning and building regulations. In the United Kingdom, by contrast, access remains a building design issue, and is only partly addressed in the new Disability Discrimination Act. New Zealand's main access regulation (in the BA) mirrors the British approach, and similarly reflects a liberalised attitude to enforcement.

Does the dual emphasis of the New Zealand approach make for a stronger regulatory regime than those existing in either the United States or Britain? This question cannot be answered here and must be the subject of a broader comparative study of national access regimes. However, the following analysis does raise serious doubts about the effectiveness of either regulatory approach in New Zealand.

The Dunedin case study

The context

Dunedin is New Zealand's fifth largest urban area with a population of 119,612 in 1996. Local government is provided by the Dunedin City Council (DCC), which, among other things, has primary responsibility for land use planning and building control in the Dunedin urban area. Since the late 1970s, Dunedin has experienced almost stagnant population growth and a protracted economic downturn (Horton, 1996; Welch, 1996). This contrasts with New Zealand's larger cities which, over the same period, have fared better economically and have generally grown faster in population. Economic stagnation has seriously eroded the revenue base of the DCC (Horton, 1996).

The 1988–89 reform of local government legislation sought to transform councils into entrepreneurial units that could compete for investment within a national economic context changed radically by the imperatives of capital mobility and globalisation (Pawson, 1996; Welch, 1996). This shift in governance paralleled similar changes in North America and Europe, where local states have progressively been transformed into 'promotional agencies' seeking to lure scarce, highly mobile investment capital (Harvey, 1989b; Logan and Molotch, 1987). A defining feature of this 'urban entrepreneurialism' has been the packaging for sale of cities as commodities which can add value to investors by offering 'business-friendly' (i.e., low transaction cost) environments.

Presiding over a declining regional economy, the DCC embraced the new structural changes in the hope that an entrepreneurial focus on 'corporate' (i.e., revenue generating) objectives might stimulate a revitalisation of investment – especially building construction – in the city. Indeed, Pawson (1996: 291) maintains that Dunedin provided 'the earliest example of entrepreneurialism in urban governance in New Zealand'. Welch argues that in the mid-1980s the DCC began to focus on two corporate roles, 'as co-ordinator and facilitator of economic development activity, and as generator of a positive city image' (1996: 292).

By the late 1980s, therefore, 'economic revitalisation' had emerged as an overriding administrative and political concern of the DCC. This policy emphasis assumed that both revenue generating activities and the minimisation of transaction costs to business must be given priority if other 'corporate' goals and programmes were to be afforded. In the past decade, the DCC voiced the new 'spirit of entrepreneurialism' through boosterist rhetoric, involving national advertising campaigns that sought to emphasise the city as a 'business-friendly' site for investment. None the less, the bright rhetoric and campaign slogans have been overlaid on enduring popular fears of continued economic stagnation and social decline.

Access politics

In recent years, Dunedin disability advocacy groups have argued that local government has failed to enforce the building legislation's accessibility standards. In November 1993, for example, the local Disability Information Service reported the frustration of Dunedin's disabled community with access control enforcement in the city. The Service accused the DCC of failing to administer the building code adequately, describing the council's attempts at enforcement as 'superficial window dressing' (*Dunedin Star Weekender*, 21 November 1993). Elsewhere, Dunedin disability activists argued that the city's local government had neglected its accessibility policy responsibilities by under-resourcing its building standards inspectorate (*Otago Daily Times*, 11 September 1994: 5). As one prominent member of the Dunedin branch of the Disabled Persons' Assembly (DPA) remarked: 'There is no point in having the best legislation on building accessibility if you do not enforce it' (*Dunedin Star Midweek*, 10 November 1993: 1).

In late 1993, after some press exposure of activists' complaints, the city council's building control manager publicly admitted that 'resources are being stretched'. He excused official inaction on access issues by remarking that 'the requirements [of the Building Act] are being enforced *as far as is reasonably practicable*' (*Dunedin Star Midweek*, 10 November 1993: 1) (emphasis added). The manager further remarked that the council was doing the best it could, 'with only one senior building inspector on the road at any one time' (ibid.). In a further admission, this same officer acknowledged that the council had not required a particular commercial establishment to install a lift during a major refit, although the building legislation may have required this. He then attempted to reassure the city's disability community with the observation that 'this place can still be accessed by people with disabilities who are not confined to a wheelchair' (ibid.), thereby demonstrating a highly selective notion of disability which conflicted with the inclusive aim of the legislation.

Later, in an exchange of correspondence with the Disabilities Co-ordinator at Dunedin's major tertiary institution, the University of Otago, the building control manager observed that:

> it is agreed that if all buildings comply 100% with NZS 4121 our city would be a better place to live in. However, because a number of our buildings are existing buildings and there are foundation design constraints ... it is not always possible to have [this].[6]

In other words, the DCC had decided that it would not fully enforce the access legislation because the nature of the existing built environment made this onerous. In fact, the legislation allows local authorities no such discretion in the application of access controls.

And if disabled people were unhappy with the DCC's (non)enforcement of the access legislation, the manager advised simply that 'The alternative is not to patronise the establishments that don't provide the necessary facilities for all people'. This suggested use of 'consumer power' in place of regulatory enforcement rests on the absurd neo-classical assumption that the 'invisible hand' of the market can supplant laws which are designed to address social discrimination. New Zealand's disabled people suffer significant economic impoverishment (see Cahill, 1991) and therefore cannot be expected to counter the discrimination they experience through the manipulation of consumption patterns. In a revealing comment in the same letter, the manager claimed that DCC 'inspectors police the code to the point of being a bureaucrat in some situations'. Moreover, 'All owners are not sympathetic to the requirements of the disabled code (sic) and some projects have not gone ahead because of those requirements'.

The comment that the DCC was overbearing in its enforcement of access standards seems hard to accept in view of the agreed under-resourcing of the building control inspectorate. However, the manager's observations that some owners resented the access controls and that these regulations had (supposedly) stopped some developments may reveal some of the broader tensions that are undermining both compliance with, and enforcement of, the building code. The feeling that access controls had prevented some developments doubtless reso-nated uncomfortably with the DCC's concern about the city's declining economy and its desire to promote Dunedin as a 'business-friendly' environment.

During 1994 and 1995, a series of articles in the local press (e.g., *Otago Daily Times*, 11 February 1994: 5; *Dunedin Star Midweek*, 12 October 1994: 3) made it obvious that the access issue was continuing to anger the city's disability com-munity. It was in this still simmering political context that the interviews with key informants were held during 1995.

The views of interest groups

The interviews with the seven leaders of disability advocacy groups (all of whom were disabled) tapped a broad cross-section of opinion; informants related their own experiences of access and also relayed the views of members and friends that had been reported to them. Several themes emerged consistently during the interviews with the group leaders.

First, all informants felt that the Building Act was not as effective as it might be, as a result of problems both in the legislation's design and in its implementa-tion by councils, including the DCC (although opinions varied about the per-formance of the latter). A number of deficiencies in the legislation were identified, including the vagueness and complexity of wording, and the failure to recognise the needs of specific groups, such as scooter users. An important further concern was the hesitancy of the Building Act's wording, especially the 'reasonably prac-ticable' clauses which several informants identified as a major loophole in the legislation. Related to this was the criticism raised by some informants of the opportunities for exemption from compliance which the legislation provided:

> exemptions can be granted so people don't have to meet the Act or stand-ards and as I understand it is very easy [in Dunedin] to get exemptions. [It] makes a mockery of the whole system.

This criticism raised the issues of implementation and the interpretation of the legislation's often vague and permissive wording by local government. Non-enforcement of the legislation was seen as a critical problem – all informants were personally aware of specific instances of both non-compliance and non-enforcement within Dunedin. Many related that new or altered commercial establishments, such as bars, pubs, restaurants and shops had failed to provide accessible toilet facilities, thus making these facilities unusable for many disabled people. The President of the University Disability Action Group related one such instance involving the construction of new toilets in the students' union building:

> I know the university was told with the new toilet in the union that there was a problem with complying with the Act … as soon as it was pointed out that there was a problem, they were told [by the DCC] 'but you don't have to if you don't want to. We will give you a letter or whatever to say it is too hard.'

The Disabilities Co-ordinator at the university recalled the same incident and her reaction to it:

> even at varsity here they had a toilet which didn't meet the code and the DCC said 'we'll give you an exemption'. I went off my face. I said we of all people should have got it right. You accept an exemption and I'll go public. We still haven't fixed the problem. The toilet is useable, you can get in. The problem is if you open the door – the lock isn't right – you get exposed in all your glory to everybody else.

This same informant related a further 'horrific' case involving a restaurant which managed to get DCC approval for the removal of a ramp:

> they had a ramp which was pretty steep, but you could still get up [it], and they asked if they could put a flight of stairs in because they said the ramp … and the [existing] stairs were unstable. To make them stable they had to take out part of the ramp so there is no wheelchair access … We know damn well it is just because they want to have that area artistically done. That took all our access away. There are so few restaurants in town where you can go with access and few of those have got wheelchair toilets.

Several reasons were given for both the perceived failure of the DCC to enforce the legislation and the unwillingness of the private sector to comply with it. Apart from problems with the legislation itself, these perceptions of deficiencies in the DCC included poor staff training, patronising or discriminatory staff attitudes, a lack of consultation with disability advocacy groups, and the under-resourcing (in terms both of finance and of staff) of the building inspectorate. One informant who had inside knowledge of the DCC building control procedures suggested that the council was taking the legislation's performance standards approach too far by 'rubber stamping' plans at application stage and then only monitoring compliance after construction when it may be too late to amend breaches of the code:

> when plans come in for permit they tend to cross over them and it is not until the alteration has been developed or the new building has gone up that the mistakes come in … they must be running under terrific pressure … it's limited staff, lack of money.

In terms of the private sector, several informants saw the concern for profit as a reason for non-compliance:

> It comes back to how the person sees themselves providing a service to the public. Is it just the dollars that matter or is it the letter of the law?..can we get around the letter of the law?..unfortunately, money is the big thing in a lot of it. If there's a cheapest way to be done, do it.

Most informants thought that the council's performance in applying the legislation could be improved through better resourcing and staff training for its building inspectorate. In addition, one interviewee proposed that the Building Act be amended to make the issue of exemptions subject to an open decision process, involving public notification of exemption proposals and third party appeal rights.

The Dunedin manager of a national disability services provider confirmed many of the concerns raised by advocates. The manager also intimated his belief that major players in the land economy were able to dictate the terms of compliance with the access controls:

> there are some issues of financial pressure ... of a developer wanting to build a building and then who is going to have the strength within the [DCC] to argue that x numbers of dollars have to be spent on making this building accessible. The developer's attitude is: 'you are talking about a very small percentage of the population and I am not going to get a return for dollars'.

For their part, the three council informants (two building control officers and the building control manager) evinced a far more positive view of the success of access controls in Dunedin. Perhaps this is not surprising given the officers' professional attachment to this domain of regulatory practice. Moreover, their relatively sanguine attitudes might also be considered in the context of the many critical comments made by disability advocates about the council officers' seeming misapprehension of how disabled people actually use the built environment. (It was for this reason that many advocates argued for a greater role by disabled people in the monitoring of compliance by building owners.)

The two building officers were particularly positive about the effectiveness of the legislation. Neither thought there was any particular problem with either the expression of the legislation or the forms in which it was made available to the public. Neither officer saw resource constraints as an issue. As one put it, 'We are always busy but that is just a fact of life; but it is not to the point of [us] having to skim over [code standards]'. This contrasted with the manager's opinion:

> No we haven't got enough personnel. We can get people in – temporary things like that. It's actually inspectors that we need – people on the ground ... We could do with ... more ... just to keep up with what we've got.[7]

Both of the control officers thought that non-compliance was rare but said it did occasionally come to their notice:

> Occasionally you do get somebody trying to do it. They are usually a small developer who perhaps doesn't have the financial resources and who is working within the confines of an existing building ... We do have a wee bit of fun with that one clause trying to make it [compliance] 'as close as reasonably possible'.

However, two facts emerged from all three officers' comments that bear consideration here. First, all informants admitted in various ways that the council could never be certain that building owners and developers always complied with the access regulations. The problem was particularly acute with 'minor' alterations, many of which were easily concealed by owners. In such cases, the council relied on members of the public to inform them of non-compliance, as one control officer pointed out: 'If we get a complaint we would act on it but it is the building owner's responsibility to comply with the Act and therefore make sure [the alteration] is correct'. The other officer admitted: 'There are a few cowboys out there who do a bit of work ... without informing us. Somewhere down the track somebody is answerable for that work that has been done.'

Second, all three informants gave evidence that 'compromise' was a key feature of the council's enforcement practice. As one control officer remarked:

> sometimes we have to make a slight compensation for [the cost of compliance] ... it comes down to ... how much the alteration will cost ... In this council we have reasonable people on both sides that won't let anything go willy nilly but at the same time they are not dogmatic bureaucrats that are sitting there saying, 'you've got to be exact'.

The manager further explained this practice of 'second best':

> It's existing buildings that can be the problem. But we can get around it. It may not be 100 percent ... we do our best and the architects and designers, they do their best to make it right. Sometimes it is just impossible and you've got to accept second best.

Is the invisibility of 'minor' alterations a significant issue? Does the council's 'slight' flexibility and compromise in enforcement practice really affect access for disabled people? Given the testimonies of disability advocates, it appears that the answer to both these questions is *yes*. It must be remembered that even seemingly minor infractions of the access code – such as unsuitable toilet provision – can cause distress to disabled people and may even make a building inaccessible to them. Clearly, as disability activists were at pains to point out, council officers had no real appreciation of what it was actually like to be a disabled user of public buildings in Dunedin. This point was exemplified by the rather dismissive reference made by the manager to the restaurant staircase alteration which one disability advocate had described as a 'horrific' case (see above):

> that's an interesting one ... we have been criticised for that because there is [now] no ramped entrance for wheelchair people but you can actually get into it I believe from Albany Street at the back ... but unless people knew their way around Dunedin they are not going to know how to get to Albany Street.

A control officer commented on how the 'reasonably practicable' clauses were used as a basis for flexible regulation:

> Usually [the issue] comes about by a ramp access. It might be a little steeper than it should be ... so we accept perhaps something that isn't quite right with the provision that the people operating the building would know that if a person turns up in a wheelchair, that there is a sign up to point out [that this person must] get a staff member to come and help them.

In fact this sort of 'compromise' is not provided for in the Building Act – the stricture that disabled people must make assisted entry to a building, when staff are available, is both unjust and *ultra vires*. One disability advocate had spoken of the problem of assisted access and had related the determination of disabled people to oppose such demeaning 'compromise solutions'. It was noteworthy that all three officers saw problems with monitoring compliance after a building consent is given. It was pointed out that approved access features are often later converted to other uses; it was known, for example, that accessible toilets were sometimes after construction converted to uses such as staff cloakrooms or storage cupboards.

All officers admitted, to varying degrees, that building owners occasionally expressed resentment at the access controls. As the manager observed, 'We've had developers moan about us wanting access improved'. The reasons for complaints included both the fact that council was seen as 'too strict' in its application of the building code and that compliance imposed unnecessary transaction costs (consent fees) on owners: 'Everybody would complain about the fees and we are aware that some builders drop the value of work down to get into the lower (fee) category.'

In summary, there was a marked cleavage between the views of disability advocates and those of DCC building control staff. The former group were unanimous in their criticism of the council's enforcement of the building code (although the strength of criticism varied from mild to bitter). Advocates identified many breaches of the access code by public building owners and all felt that the council under-resourced its building control operations. Council officers, by contrast, were relatively positive when speaking generally about the effectiveness of the code, although at the detailed level their remarks revealed several areas where enforcement may be seriously deficient.

Reasons for non-enforcement

How are the findings from the case study to be understood in terms of the wider institutional and theoretical contexts which define access issues? First, it appears that regulatory practice in Dunedin is sealed off from a critical evaluation source: the opinions and experiences of disabled people. Council officers involved with the Building Act's application demonstrated a general satisfaction with the effectiveness of enforcement, in spite of both some questionable flexibility in interpretations of the legislation, and the publicly expressed dissatisfaction of disabled people with the council's performance.

Could enforcement be improved through a more effective liaison between the DCC and Dunedin's disability community? Perhaps, but there appear to be limits to this strategy. The Building Control Manager had in fact been a member of a local advocacy group's access committee, but had gradually become an irregular attender of meetings prior to the time of the interviews. This is not surprising in that the manager undertook this role outside work time on a voluntary basis and it was perhaps inevitable that the evident pressures of a difficult job would eventually reduce his capacity for informal liaison. More effective liaison would have to be structured into work time, and would doubtless have resource implications for the council (indeed, several advocates argued that the council should be paying them for the *ad hoc* consultations they had been providing for several years.) Given the DCC's straitened finances, it is most unlikely that resources would be devoted to liaison.

In Britain, Thomas (1992) has pointed to the potential of formally constituted local liaison groups to assist councils to implement access controls. Without public resourcing, however, such liaison mechanisms are likely to fail: overall, disabled people in New Zealand, as in Britain, are economically disadvantaged and have little organic capacity to provide support for such time-consuming activities. This point was exemplified in Dunedin in February 1996 when the local branch of the peak national disability advocacy organisation collapsed. An immediate past president of the group blamed the group's demise on 'overwork and under-resourcing' and felt that 'unpaid consultancies' with public sector organisations had simply placed too much stress on the volunteer members (Mackay, pers.comm., 1996).[8]

Have structural factors – such as local state growth politics and the commodity land economy – affected the enforcement of building regulations in Dunedin? This is a difficult question to answer given the nature of the data derived for the case study. Certainly, there are indications that many owners and developers of land see the costs of compliance with access controls as irksome, and perhaps as an unnecessary distortion of market signals. Given the general impoverishment, underemployment and socio-spatial exclusion endured by disabled people, it would not seem surprising if property developers saw no reason to spend money on opening their facilities to this marginal consumption group.

There was no direct evidence in the data that the council was consciously weakening its building regulation in order to ensure that Dunedin was seen nationally as a business-friendly investment environment. In terms of transaction costs, at least, there was no evidence that the council was providing cheap regulation – fees for building consents were hardly inconsiderable (they ranged from NZ$210 to NZ$2,250). As the building manager admitted, these fees were certainly of concern to developers and may have contributed to non-compliance. However, while regulation was not cheap, it was also not rigorously applied. The fact that compromise and 'flexibility' were key features of building control 'enforcement' may indicate that the ideology of growth politics had permeated the council's regulatory practices in subtle ways. The suggestive, if inconclusive, nature of the research results in this respect indicates the object for a more comprehensive study of access regulation.

Other broad ideological factors may also be limiting the effectiveness of the building legislation to improve disability access in Dunedin and elsewhere. As mentioned earlier, the legislation is a key example of the new, 'flexible' regulation that has emerged since the ascendancy of neo-liberal politics in New Zealand from 1984. The performance standards approach of the Building Act embodies the neo-liberal concern for flexible, economically efficient regulation which achieves clear objectives while none the less minimising transaction costs to the private sector. There were indications in the case study that the performance standard approach of the building legislation was hindering rigorous enforcement. In particular, the observed tendency of council officers to subject only the outcomes of building development to detailed scrutiny emerged as a potential problem area. By this stage, the physical and economic fixity of investments are likely to make inspectors unwilling to order significant corrections to code breaches, especially in a policy domain defined by compromise and flexible enforcement.

Conceivably, performance flexibility might improve the code's effectiveness if building owners are willing to observe the spirit of the legislation and seek innovative ways to comply with standards. This would also require significant staff resourcing to ensure that all development proposals and outcomes are monitored on a case-by-case basis (an inevitable consequence of flexible, performance-based controls). However, from the case study data, it appears that neither of these criteria can be assumed for Dunedin.

Finally, the indeterminacy of the building legislation's strictures, centring on the 'reasonably practicable' clauses, is clearly a source of confusion, even evasion, in regulatory compliance. Thomas (1992) is also critical of similar 'reasonable compliance' clauses in British access legislation. The British legislation:

> with its references to practicality and reasonableness ... emphasises ... optimum solutions in situations involving competing needs, or interests. Thus might a fundamental right to an independent and dignified life be reduced to an 'interest' to be balanced against [other] requirements.
>
> (Thomas, 1992: 25)

It is well here to recall one such 'reasonable' solution to building owners' design problems in Dunedin involving the provision of demeaning (and very possibly illegal) 'assisted entry' access for disabled people. That such solutions are judged 'reasonable' can only be explained if one is prepared to accept that economic and bureaucratic imperatives take precedence over human rights in enforcement practice.

A national problem?

Are the findings of the Dunedin study generalisable? Are there systemic problems with the enforcement of access controls in New Zealand which suggest the influence of structural limitations? The scarcity of any further primary or secondary evidence on access law policy practice means that these questions cannot be answered at present with any certainty.

However, there are some fragments of secondary evidence which suggest at least that there are problems with the enforcement of, and compliance with, access controls in other local government contexts. For instance, a recent spot survey of public buildings by one North Island council revealed significant non-compliance with building controls, including access standards.[9] In this survey, nearly 40 per cent of buildings were found to be in violation of the access code. A separate set of surveys carried out in New Zealand's largest city, Auckland, in 1993 indicated similar compliance problems in several of the city's major hotels (*New Zealand Disabled*, December 1993: 6).

A rights based approach?

Can the Human Rights Act address the problems which may be emerging from non-enforcement of the Building Act? On the face of it, the HRA has considerable potential for improving the accessibility of New Zealand's cities. For instance, the legislation extends the notion of access to transport modes, employment practices, service delivery and accommodation. This potential was tested in 1994 when disability advocacy groups lodged a series of complaints with the Human Rights Commission (HRC) alleging that regional councils and private transport operators were failing to provide accessible bus systems in New Zealand (*Otago Daily Times*, 9 August 1994: 7). The complaint centred on the decision of public and private transport providers to purchase buses which disability groups considered inaccessible. In response, the transport manager for Wellington Regional Council (the subject of one complaint), observed that he 'was concerned about costs being imposed on ratepayers to benefit a group of people of unknown size' (ibid.). This insinuation that disabled people were a shadowy, but troublesome, minority in New Zealand was an extraordinary distortion of the truth. In fact, some 37 per cent of New Zealanders are estimated to have at least one form of impairment (Statistics New Zealand, 1993). Moreover, the transport accessibility issue is of immediate concern to a range of other social groups with 'special' mobility needs, including the elderly, pregnant women and users of prams (Wrightson, 1989). In March 1995, the chairperson of one regional public transport provider warned that an HRC ruling in favour of universal access might 'kill' the national bus system:

> If we accept [such a] decision and put the demands into place we would not have a public transport system; we just wouldn't be able to afford it ... if we adhere to the Human Rights Act it will be at the peril of our efficient and well-used system.
>
> (*Otago Daily Times*, 8 March 1995: 13)

The irony in describing an exclusionary bus system as 'public' was evidently lost on this commentator. Opposition from transport operators towards an inclusionary ruling by the HRC deepened during 1995. Reflecting this mood, the executive director of the Bus and Coach Association urged transport operators in early 1995 to 'take a stand against the Human Rights Act' (*Otago Daily Times*, 8 March 1995: 13).

It is difficult to say whether the HRA could be used as superior law to override and correct flaws in the implementation of the building legislation. If recent experience of disability advocates groups on transport access is any guide, any attempt to extend the HRA to the built environment would doubtless meet strong resistance from the owners of public buildings.

Similar problems have been encountered in the United States, where many employers, building developers and public transport operators have, in varying ways, resisted implementation of the ADA. As with the New Zealand case, American public transport providers have argued that the ADA has imposed crippling financial burdens upon them. In 1995, transit authorities in several major United States cities – including Washington and Madison – were openly complaining about the ADA requirements. Transit operators maintained that they simply could not afford to comply with the law by making transport vehicles and networks more accessible. The situation was not helped by the Clinton administration's cutbacks to transit subsidies provided to cities through the Federal Transit Administration (*Washington Post*, 20–26 March 1995: 31). As Davis comments,

> The ADA is only as effective as its enforcement. But there is no federal agency to enforce the provisions of this law ... the weight of the law can only be brought to bear through a lawsuit or fear of a lawsuit. But lawsuits are costly and time-consuming, and to bring them is beyond the means of most people with disabilities.
>
> (1995: 159–60)

Certainly the ADA has achieved major gains for disabled Americans, principally in the form of physical improvements to many public facilities. None the less, the evidence thus far suggests that the ADA has not eliminated the underlying structural causes of disablement, such as, for example, the disabling division of labour. More than four years after the passage of the ADA, the number of disabled people entering the United States workforce had not increased. Indeed, one survey – by the National Organization on Disabilities – indicated a deterioration in the general workforce status of disabled people since the enactment of the ADA, finding that the proportion of disabled adults in full or part time employment had declined from 33 per cent in 1986 to 31 per cent in 1994 (*New York Times*, 23 October 1994: 18).

Dorn (1994: 104) is convinced that the ADA represents a 'revitalization of citizenship for disabled people', but this claim is open to challenge if citizenship means anything more than a set of legally enshrined rights.[10] Clearly, these rights conflict in many circumstances with the deeper political-economic and cultural imperatives that frame contemporary capitalist societies. Rights legislation seems especially vulnerable to resistance from key cultural and economic interests in society, and also to the fluctuating commitment of states to justice for disabled people. In Australia, for example, major human rights protections for disabled people were greatly undermined in 1997 by a series of financial cutbacks and institutional changes imposed by a new neo-liberal national government (*Canberra Times*, 24 September 1997: 3).

In summary, it seems that rights legislation is perhaps a necessary, but certainly not a sufficient, precondition for achieving enabling environments for disabled people. This observation echoes the more general criticisms made previously against rights-based approaches to social justice in capitalist societies. Indeed, Marx (1977), writing in the mid-nineteenth century, poured scorn on the pretensions of rights-based notions of justice that were advanced at the time by political liberals. As he observed, the 'pure morality' of bourgeois reformism was largely rhetorical and gestural, if well meant, and did little to change the underlying socio-spatial patterns that produce inequalities in capitalist societies. While not suggesting that contemporary examples of disability rights legislation are completely ineffectual initiatives, I do argue that these approaches cannot by themselves eliminate disability oppression.

Conclusion

What has this study shown? First, a theoretically informed analysis of access regulation in a range of Western countries suggested that legislative compliance is likely to be compromised in policy settings by institutional-structural factors. Broadly, these delimiting influences include the social oppression experienced by disabled people generally, in concert with the commodified land economy which tends to shape built environments to reflect the mobility needs of 'average' consumers and workers (Hahn, 1986). In addition, some observers of policy practice in Britain and the United States have argued that the now established emphases on deregulation and public entrepreneurialism in local governance have undermined the application of access regulations.

It was clear from the Dunedin case study that access regulations were failing to address the mobility needs of disabled people in that city. Many building owners were failing to comply fully with the access legislation and it was evident that these controls were frequently seen by businesspeople as a cost burden which was to be avoided if possible. It was also evident, in spite of the stated opinions of council officials, that access controls were not being fully enforced in Dunedin. While the study established these problems of non-compliance and

non-enforcement as facts, neither their scale nor their origins were fully eluci-
dated. Both the adequacy of the access legislation and the effectiveness of its
implementation are at issue, but further study is needed in order to clarify the
relative significance of both potential sources of non-enforcement.

The study intimated that the origins of observed problems with access regulation in Dunedin may, in part, lie embedded in wider socio-economic relationships. It was certainly evident, for example, that there were frictions between the
local land economy and the access regulations. Also, the general influence of neo-
liberal politics on the Dunedin City Council is not in doubt, but it was not
possible in the study to establish clearly whether subsidiary ideologies – notably,
public entrepreneurialism, cost cutting and performance regulation – were erod-
ing the effectiveness of access controls.

As I explained, there are strong theoretical reasons and limited empirical grounds
to believe that the problems observed in Dunedin with access regulation are
common to other regional and national contexts. However, only a broader, empirically informed investigation of policy practice can establish whether this inti-
mation is a fact. If the problems relating to access regulation are to be properly
understood, it will be necessary to isolate through comparative empirical analysis
the reasons for any observed non-enforcement and non-compliance. Thus, there
is a clear need for similar studies in other Western countries to facilitate international comparisons of access enforcement and compliance. Such comparison would
help clarify whether the divergent national approaches to access regulation actually lead to distinct – i.e., more or less effective – compliance outcomes. These
analyses must seek to establish to what extent such problems are structural (e.g.,
related to political-economic factors, such as national growth politics), and/or in
what ways are they contingent (i.e., place specific anomalies in implementation
and observance of laws).

The empirical clarification of these issues would in turn contribute to the political task of making access regulations more effective in Western cities. In particular, such research would help advocacy groups to target their political energies
more accurately on the structures, practices and contexts that most inhibit the
realisation of accessibility policy aims. Most importantly, perhaps, any such investigation must privilege the opinions and experiences of disabled people, for whom
the seemingly innocuous terms 'access' and 'exclusion' can literally mean (social)
life and death.

10 Towards an enabling geography

Introduction

This concluding chapter has two aims: first, to recapitulate the theoretical and empirical arguments of the book; and second, to consider the ways in which the discipline of Geography might play an enabling role in disabled people's struggles for justice and respect. Although I focus here on the potential for an 'enabling Geography', I think that the remarks made apply readily to the other spatial disciplines that have at times been addressed in the book. Simply put, disablement is a profoundly spatial experience, meaning something that is lived and produced at every imaginable scale, ranging from a chair or stairway through to the macro policy realms that constitute state institutional practice. As such, disability spills beyond what has traditionally been codified as the bounds of geographic enquiry. As I have tried to show, it is a *spatial perspective*, rather than simply geographical knowledge, that has been lacking from established debates on disability within the social sciences and the humanities. Architects, urban planners and environmentalists can and must contribute to a wider debate on the production and transformation of disabling social spaces.

Summary of the arguments

This book presented a multi-level conceptualisation of disability, sourced in the method of historical geographical analysis. The enquiry was framed by a broad theorisation of the historical-geography of disability, centring upon the experience of physically disabled people in capitalist societies. This theoretical framework – embodied materialism – was constructed through analysis of the key ontological issues surrounding disability; namely the social interrelations of time, space and embodiment. In Chapters 2 and 3, I argued against various reductionist theories which assume disability to be a simple reflection of natural or social forces. Instead, I posited a historically and geographically informed model of disability that recognises the material reality of impairment while stressing the specific ways in which this form of embodiment is socialised in different times and places. My historical-geographical theory of embodiment was painted in the broadest of brushstrokes, intended as an abstract *materialist* framework that could both guide and accommodate a variety of conceptual-empirical studies of disability. I concluded Part I of the book by calling for new historical-empirical studies of disability that could help to explain the ideologies and practices that surround the socio-spatial construction of impairment in contemporary capitalist societies.

In the second part of the book, I made my own contribution to this task through a set of historical-geographical studies of disability in different, though closely linked, societies. I began the studies by first addressing in Chapter 4 the most pressing historiographical issues for both case studies. This discussion briefly analysed how the constitutive relationships of a mode of production condition the socio-spatial construction of disability. The studies of medieval England and the industrial city exposed the different ways in which impairment was socialised within the feudal and capitalist modes of production. The central conclusion was that capitalism in the past has tended to *desocialise* disabled people by producing landscapes of exclusion (cf. Sibley, 1995). The use of the verb *condition* is critical here – while the enquiries revealed the desocialisation of impairment in capitalist societies, they also exposed some of the many ways in which disabled people have resisted and transformed these oppressive structural tendencies.

In the third part of the book, the empirical-analytical focus shifted to contemporary Western capitalist societies. Chapter 7 was a general review of the urban context of disablement in Western societies, concluding with a political-ethical framework – the *enabling environment* – for analysing contemporary social spaces of disability. In Chapters 8 and 9 I examined two state policy realms that have great consequence for disabled people: community care and accessibility regulation. My analyses showed that both realms are failing to provide enabling environments in Western cities. There are several factors that are undermining the enabling potential of these policy environments, including entrenched social attitudes and practices, along with core political-economic relations. I argued in both studies that the neo-liberal political-economic agenda, in particular, has grave consequences for marginalised social groups, including disabled people. Neo-liberalism has the demonstrated ability to erode and/or distort progressive social and environmental policies in ways that worsen the injustices experienced by disabled people.

There were many social spaces of disability that I could have addressed in this part of the book – the case study choices simply reflected my personal and professional interests. Geographers, and other social scientists, have already provided critical accounts of other policy spheres, including health, leisure, housing and transport. I hope that there will, in time, be a flourishing geographical discussion on public policies for disabled people, that can, in turn, contribute to and refine the debates in disability studies and within the social sciences generally.

In summary, I would like to think that this book has contributed theoretically and empirically to the awakening interest in disability issues within Anglophonic Geography. And I am already aware that my contribution will merely be one of many in a new, and vigorous, area of geographical enquiry. Not a month passes, it seems, without some new professional or research initiative by geographers interested in disability. Very quickly, it appears, the discipline is moving from indifference to real engagement on this issue.

By using the term 'contribution', I do not mean to suggest that the theoretical arguments and empirical studies in this volume will be received with universal favour. Indeed, I set out in these essays to oppose certain theoretical tendencies – notably, various forms of reductionism – that have emerged in the social scientific accounts of disability, including some in Geography. I therefore expect that the arguments of the book will join a rapidly expanding and hotly contested debate on the relationship between space and disability.

New geographies of disability

In the introductory chapter I argued that it was the task of a social movement to articulate and effect the profound cultural, political and economic changes that will be required to eliminate disability oppression. As indicated in Chapter 7, disabled people in recent decades have formed increasingly influential social movements across Western countries to achieve this aim. Great gains have been made, and yet large obstacles to progress remain while new ones emerge with depressing frequency as social disadvantage thrives in an era of 'market triumphalism'. There is much to do.

What role can Geography and geographers play in the continuing struggles of disabled people? A *comprehensive* answer to this difficult question would take a book in itself, and, in any case, I am not sure that it is appropriate for a non-disabled person, such as myself, to make this sort of political gesture. Drake (1997), musing on this very issue, has suggested that non-disabled people might properly contribute ideas and opinions to disability discussions, but should not engage in decisive acts that attempt to control the political agenda of disability movements. Obviously the distinction between a 'contribution' and an 'intervention' is going to be difficult to discern in many contexts. Drake offers one useful rule of thumb, arguing that

> while it may be acceptable for 'non-disabled' people *to join with* disabled people *to lobby for* anti-discrimination legislation, it is in my view unacceptable for them to lobby on behalf of disabled people.
>
> (1997: 644) (original emphasis)

Taking this wise counsel, I deem it inappropriate for me to present here a comprehensive manifesto for an enabling Geography. This would cross the line separating contribution from intervention. In any case, I confess that I have no idea where Geography might ultimately take us in the struggle against disability oppression. However, I do believe that this issue will be clarified as the discipline confronts the question of disability with the sort of seriousness that is now rightly given to other forms of social cleavage, such as class, gender, race, age and sexuality. Moreover, for this 'confrontation' with disability to occur, it will be necessary for the discipline to remove the institutional barriers that have long isolated Geography from disabled people and their social movements.

So, to conclude this book, I offer some thoughts on how Geography might remove some of the barriers that the discipline has erected between itself and disabled people. More specifically, I will speculate on the sorts of shifts in theoretical and practical research agendas that may be needed if Geography is to play a role in the larger, emancipatory struggles of disabled people. My approach is premised on a belief that the enabling potential of Geography will not be fully known until these barriers are swept away.

Research themes

In Chapters 1 and 2, I pointed to the new spatial studies of disability emerging from a variety of national contexts within Anglophone Geography. However, it is true that most of this new work on disability has been produced by a rather small coterie of social and cultural geographers. There are still systematic areas of geographical enquiry – notably Economic, Urban and Historical Geographies – that have thus far shown little interest in disability issues. In what follows I briefly consider how disability might be extended to these other sub-disciplinary concerns.

The production of urban space

In the Urban Geography literatures, much work has been done in recent decades to explain the social production of space, concentrating on how sets of structural and institutional forces condition the formation of cities and other urban spaces (Johnston *et al.*, 1994). The many analyses which have contributed to this understanding have emphasised different aspects of this complex phenomenon, ranging from political economies of how capital shapes cities (e.g., Harvey, 1989a; Smith, 1984), through institutional geographies examining the regulation of space and the spaces of regulation (e.g., Badcock, 1984; Pinch, 1997), to recursive studies that link biographies and places (e.g., Rowe and Wolch, 1990).

But in this significant area of geographical enquiry, one struggles to find any recognition, let alone analysis, of a critical aspect of produced space – the inaccessibility of most cities to a substantial number of their inhabitants, including disabled people. Most major urban geographic teaching texts overlook the issues of disability and accessibility. It is ironic that the one major theoretical study of accessibility to emerge during the 1980s (Hahn, 1986) was forwarded by a political scientist, not a geographer. As noted in Chapters 2 and 9, there has recently been some attention given to the question of accessibility by urban social geographers, such as Imrie (1996a) and Vujakovic and Matthews (1992, 1994). To date, however, these analyses have not managed to register accessibility as a major concern for Urban Geography.

As I have attempted to show in Chapter 9, the question of access regulation goes straight to the heart of theoretical debates on the production of space in advanced capitalist societies. Accessibility is a profoundly important dimension of produced space which is sourced in the ensemble of political-economic and cultural dynamics that shape built environments. As I have already argued (see Chapter 9), much of the contemporary access literature assumes a rather simple and undynamic view of inaccessibility, as if it were the 'accidental' outcome of thoughtless actions by countless individual bureaucrats, building owners and transport operators. Urban Geography, through the production of space perspective, could do much to enrich the rather impoverished understanding of inaccessibility which pervades other disciplines and public policy realms.

As I noted, there are signs that this enabling power is awakening within the discipline, most especially in Imrie's (1996a) landmark study and also the grounded work – directly involving disabled people – of Vujakovic and Mathews (1992, 1994). These analyses have all explored different aspects of the production of inaccessible space, ranging from the mobility experiences and strategies of disabled people to the role played by regulatory and political economic forces in determining levels of urban accessibility. There is much in these new analyses of disability that could contribute to the core teaching and research concerns of Urban Geography.

The economic geography of disability

Geography is yet to consider seriously the economic aspects of disability. One searches without result the major texts and journals dealing with Economic and Industrial Geography for any mention of disability in the analyses contained therein (Hall, 1994, 1997). The influential *Dictionary of Human Geography* (Johnston *et al.*, 1994: 147) defines the field of Economic Geography as 'A geography of people's struggle to make a living'. Is it not then extraordinary that economic geographers have ignored a sizeable social group whose very existence is overshadowed by profound material struggle?

Feminists in recent years have criticised Economic Geography for ignoring non-class cleavages – notably, gender and sexuality – in the social relations that underpin economic activity in capitalist societies (e.g., Rose, 1993). The same argument can be made for disability, a social identity which, as I have shown in Part III of this book, is characterised by a specific set of economic realities, including poverty, labour market exclusion, welfare dependence, and low pay. Although these conditions resonate with those that commonly define the experience of many women in capitalist societies, disability and gender are distinct economic identities. Although, for example, biology-physiology can be a source of labour power devaluation for both women and disabled people, the latter experience labour markets in highly specific ways (Abberley, 1997).[1] One important geographical difference between the economic experiences of non-disabled women and those of disabled people is the exclusion of many of the latter in special industrial realms, frequently known as 'sheltered workshops' (Alcock, 1993).

As I argued in Chapter 6, the sheltered workshop emerged within Western cities in the early twentieth century as a 'humane' solution to the problem posed by the pool of disabled labour power that industry largely refused to absorb (Ronalds, 1990). However, it was also noted (Chapter 7) that sheltered work-shops are sites both of marginalisation and of exploitation for workers within the economies of contemporary cities. Sheltered workshops promote the exclusion of disabled people from mainstream employment settings and also frequently subject their workers to low pay and poor labour conditions. None the less, the problems facing disabled workers in sheltered work settings have been poorly documented in most countries. There is a pressing need for analyses which can assess both the internal labour regimes of sheltered workshops and the more general role of these distinctive sites of production within the broader industrial landscape. In terms of the latter, exogenous consideration, it could be hypoth-esised that these cellular low pay regimes might have aggregate or at least sectoral effects on the rate and type of technological change within certain labour inten-sive industries (e.g., packaging) (see Oliver, 1991).

Moreover, these geographic analyses of disability and labour markets could be extended in a comparison of national and regional policy realms. Oliver (1993: 52) has stated that, in respect of disability employment policy, there 'are virtually no attempts in modern industrial societies that are targeted at the social organi-sation of work, at the demand side of labour'. There is, of course, the important exception of Germany, where federal laws mandate for most firms that a certain proportion of their employees be disabled (Bundesministerium für Arbeit und Sozialordnung, 1997). Though not without its problems, the German approach seems to have done much to improve the employment chances of disabled people in that country (Arbeitsgemeinschaft der Deutschen Hauptfürsorgestellen, 1995). A comparison of German and other national labour policy spheres would help support the case for demand-side regulations in Anglophonic countries.

There are, of course, other economic geographies of disability that must be written. For example, a 1970s Commission of Inquiry into poverty by the Aus-tralian government found that location was a major determinant of the levels of poverty experienced by disabled people (Gleeson, 1998). Also, two decades ago, Hugg (1979) undertook a spatial analysis of 'work disability' and poverty in the United States which showed a clear social and geographic relationship between these two variables. However, neither of these suggestive, if limited, studies seems to have gained any attention within mainstream Economic Geography. A land-scape of enquiry awaits us here.

Institutional geographies

There is now a wealth of geographic literature on human service institutions and the restructuring of this mode of social care in Western countries since the Second World War.[2] This literature, inspired by work in philosophy (notably Foucault, 1975, 1979) and social history (especially Ignatieff, 1978; Rothman, 1971) has charted historical aspects of institutional care for socially dependent people. These historical geographies (e.g., Dear and Wolch, 1987; Driver, 1993; Park, 1995; Philo, 1995, 1996) have illustrated the rise from the nineteenth-century of a 'carceral landscape' – made up of hospitals, asylums and workhouses – to which was exiled a growing estate of socially dependent groups, including orphans, the mentally ill and the elderly.

Geographers have also studied the more recent shift in most Western countries from institutional to community-based modes of social care (e.g., Bain, 1971; Dear *et al.*, 1994; Dear and Wolch, 1987; Fincher, 1978; Giggs, 1973; Kearns *et al.*, 1992; Smith and Giggs,1988; Wolpert and Wolpert, 1974). As mentioned earlier, much of the geographic work concerning deinstitutionalisation has investigated patterns of local community resistance to the neighbourhood-based care facilities that are replacing large asylums and hospitals in most Western countries (e.g., Dear and Taylor, 1982; Gleeson and Memon, 1994; Moon, 1988). In addition, geographers have shown how poorly executed public social policies have meant that many deinstitutionalised people continue to suffer poverty and socio-spatial isolation as part of the burgeoning homeless populations of many Western cities (e.g., Dear and Gleeson, 1991; Law and Wolch, 1993; Laws and Lord, 1990; Wolch and Dear, 1993).

One important criticism that can be made of the body of geographic work on institutions and community care is that it has focused upon particular social groups – notably, homeless and mentally ill people – to the exclusion of those with physical and intellectual disabilities. Allowing for confusions which arise through international variations in official terminology, a distinction must be drawn between mental illness, a disorder of the mind, and physical disablement, which relates to physiological difference. According to Park *et al.* (1998: 222), 'geographers have paid less attention to intellectual disability ... than to psychiatric disability'. The same could be said for physical disability. Both the geography of institutions and the broader field of medical-health geography have tended to focus the discipline's attention on issues of mental health, with physical and intellectual abilities receiving scant attention.[3]

Disabled people and the mentally ill have not experienced deinstitutionalisation in identical ways, and there is a need for geographic analyses that are sensitive to this difference. Such analyses should draw critical links between studies of how disabled and mentally ill people have encountered institutionalisation and its aftermath, with a view to exposing the differences and commonalities of experience across the two social groups. As Parr puts it:

> There may be an important series of connections between geographers who are becoming interested in disability, ableism, and space and the more long-standing geographical interest in mental health. These connections might include critiques directed at histories of medicalisation, and of the imposition of dualistic notions of normality and abnormality for people with psychological or physiological differences.
>
> (1997b: 438)

For example, there is great potential for new geographic work on the nursing home sector, which in recent times has emerged as an important, and frequently oppressive, institutional landscape for disabled people in Western countries, such as the United States, Australia and New Zealand (Dorn, 1994; Laws, 1993). New geographic analyses could extend the pioneering work of Laws (e.g., 1993) in the nursing homes sector, to consider both the distinctive and the overlapping experiences of disabled, mentally ill, and elderly people in this form of community care.

Dear (1992) acknowledges that popular perceptions of service dependent peoples are highly variegated and he points to a 'pecking order' of community preferences for various socially marginalised groups. At the 'highest' end of the pecking order are those social groups which arouse most fear (rational or otherwise) among communities in Western cities, including the mentally ill and people with AIDS. Dear (1992: 289) thus rightly observes that 'the intensity of NIMBY sentiments vary widely, depending upon the specific service clients'. He goes on to note that physically disabled people are probably one of the lowest placed groups on the scale of popular anxiety but he does not cite examples of geographic analyses of community reactions to physically disabled people. However, such an analysis was conducted in the United Kingdom by Moon (1988) whose study of NIMBY attitudes found that hostels for 'physically handicapped' people were seen as less 'noxious' by communities than were facilities for a variety of other social groups, including the homeless, drug users, and women seeking refuge from domestic violence. While Moon's study did not measure attitudes to facilities for the mentally ill, he is certain that 'findings would have been more negative in the case of mental illness' (1988: 213).

In view of the foregoing, it seems important to sensitise further the already considerable and insightful geographic knowledge of the NIMBY syndrome through new studies of community reactions to the establishment of care networks for 'less noxious' groups such as disabled people, the elderly and children. Work has been undertaken on community acceptance of these social groups in other areas of social science (e.g., Balukas and Baken, 1985; Berdiansky and Parker, 1977; Currie *et al.*, 1989), but none of these studies have fully developed the important spatial concepts – such as externality and distance decay[4] – that geographers have used to understand the NIMBY syndrome. Without such analytical discrimination between social groups, there is a danger that geographic knowledge will contribute to public policies which wrongly presume the existence of general social attitudes towards all groups using community support services. A practical consequence of this erroneous generalisation might be inhibited programme planning by agencies that provide residential community care for disabled people.

Research and political engagement

In this book I presented an outline for a historical-geographical analysis of disablement in capitalist societies. In common with all post-positivist perspectives, the historical-geographical approach seeks to define openly the terms of political engagement between theorist and 'subject'. The political articulation of Geography and disabled people's experiences presents a difficult ethical and intellectual dilemma for geographers that I have addressed in only a very limited way in this book.

There has been a growing awareness in recent decades among critical geographers that our work should seek an emancipatory role outside the academy, within the real social contexts where people and social groups experience everyday oppression. Chouinard (1997), as noted in Chapter 1, puts this same demand for the emerging field of disability geography, arguing for approaches that contribute to the actual political struggles of disabled people. I argued in the first chapter that published works – such as this book – can and should contribute indirectly to the struggles of disabled people, though even this role cannot be assumed and the challenge remains for geographers to make their work accessible, and therefore relevant, to disability communities. To this end, one obvious strategy is for geographers of disability to expose their work to critical scrutiny outside our own discipline, especially within the academic and political fora of disability movements. Indeed, I think it a very positive sign that the main professional forum for geographers of disability, DAGIN, has itself become increasingly drawn into the realms of disability studies and disability movements, especially within the United States.[5]

None the less, as Chouinard (1994, 1997) reminds us, geographers should not restrict themselves to the sort of formal and indirect engagement with disability politics that publications and conference papers represent. Chouinard has argued for a reconstructed and democratised radical geography which embraces the multiple political concerns of socially marginalised groups, including disabled people. For her, this new and expansive radicalism demands that geographers connect practically and politically with the experiences of marginalised social groups:

> This means putting ourselves 'on the line' as academics who will not go along with the latest 'fashion' simply because it sells, and who take very seriously the notion that 'knowledge is power'. It means as well personal decisions to put one's abilities at the disposal of groups at the margins of and outside academia. This is not taking the 'moral high ground' but simply saying that if you want to help in struggles against oppression you have to 'connect' with the trenches.
>
> (Chouinard, 1994: 5)

This then highlights the most practical, and the most challenging, demand of an enabling Geography: namely, that geographers must participate in the political struggle against the socio-spatial formations that oppress impaired people. There is a need within the discipline for a debate on how we can achieve the forms of direct engagement that Chouinard has in mind. In particular, there should be discussion on the sorts of research strategies that would be appropriate for an enabling Geography. A number of commentators – including Chouinard herself (1994, 1997), Chouinard and Grant (1995), Dorn (1994), Hall (1994) and

Kitchin (1997) – have contributed to an emerging discussion on this issue. All commentators stress the need for empowering research strategies; that is to say, engagements by geographers that contribute directly to the political needs of disability movements. In a recent consideration of this issue, Chouinard (1997) has outlined a number of enabling research methods that prioritise power-sharing. Specifically, these methods involve the transfer of technical skills and information to disabled people, and the pursuit of inclusionary research processes that de-centre the priorities of the researcher. Similarly, Kitchin advocates a 'participatory action research' (PAR) model that attempts to 'facilitate a moral geography of social action through the facilitation of studies *with and by* research subjects' (1997: 2) (original emphasis). Echoing Chouinard's 'power-sharing' approach, the PAR model 'seeks to fully integrate research subjects into the research process from ideas to data generation to analysis and interpretation to writing the final report' (ibid.).

Of course, an enabling Geography must do more than identify empowering research methods; it must also locate, and engage with, the political arenas of disabled people and their various movements. The task of engaging with disability is unavoidably challenging for researchers – it demands both that we think politically about our work and that we expose ourselves to direct political evaluation. However, the task of locating 'places of engagement' is not nearly as difficult – indeed, many of us are already situated within important domains of struggle for disabled people. In spite of the barriers to educational achievement that I discussed in Chapter 7, there are many disabled people in places of higher learning, among our students and colleagues. Hence, our own workplaces – universities, research institutes and bureaucracies – are probably the most appropriate starting places for the emancipatory engagements that Chouinard envisages. In many instances, these institutions are disabling places, presenting physical, intellectual and administrative barriers to the development of disabled students and staff (Harris *et al.*, 1995). Many such institutions have disability advocacy fora – usually attached to student organisations – that welcome involvement by non-disabled academic staff. My own participation in such groups has enriched my appreciation of disability politics – universities are in many ways microcosms of the broader arenas of struggle for disabled people. I also learned much about inclusive teaching methods, everyday access issues, and the disabling practices of educational institutions.

Golledge (1993) calls for geographical research which can enhance the ability of impaired people to cope with the experience of disability. While I appreciate his impatience with social science that offers little of value to the everyday lives of disabled people, I believe that an enabling Geography should aim to do more than simply ameliorate the effects of disablement. The historical-geographical position that I have developed in this book implies that disability can and must be opposed at a deeper socio-political level; namely, at the level of processes that create social space and thereby shape the social experience of embodiment. In particular, structures such as the commodity labour market and the capitalist land economy can be identified as critical realms of emancipatory struggle, given their importance in creating landscapes which exclude many social groups, including disabled people.

Of course, these are grand political aspirations and I think that they could only succeed ultimately as part of a broader progressive shift away from the oppressive and alienating relations that frame capitalist societies. None the less, as the various national and regional disability movements have shown, there is much that can be done in the meantime to confront the sources of disability oppression. Geographic analyses could contribute to these emancipatory movements by suggesting strategies, policies and regulations that aim to counter core disabling relations. For instance, it was pointed out earlier in this chapter that Western governments have not attempted to regulate the demand side of disabling labour markets. Rigorous comparative analysis of alternative approaches, such as the German labour law, might help to foster support for enabling laws and policies in Anglophonic (and other) countries. This sort of research requires a scale of political engagement that extends beyond the university to the level(s) occupied by disability movements that aim to influence state policies and practices. Obviously, there are many other levels of engagement that lie between the workplace and the polity, including the large variety of community struggles waged by disabled people.

Whatever our scale of engagement(s), there arises the need for a political-ethical outlook that can guide an enabling Geography. Ultimately, this ideal can only be defined by disability movements themselves. However, I believe that geographers can and should contribute to this process of political-ethical definition. In this book I have suggested that an enabling Geography requires an inclusive, but not homogenous, ideal of social justice. More specifically, I argued that this ethical ideal would have material fairness, socio-cultural respect and socio-spatial inclusion as its central political objectives. I further argued that these objectives cannot be achieved through the promulgation of universal moral standards alone. Rather, they must be won through political engagements which presume social difference and seek thereby to articulate and satisfy the variety of human needs that exist in contemporary capitalist societies. The idea that engagement is the means to enablement is echoed in a call by Chouinard and others (e.g., Hall, 1994) for geographers to contribute directly to the movements organised by disabled people themselves. Direct engagement is the most powerful way of ensuring that geographic research serves – i.e., *empowers* – disabled people and thereby avoids the tendency of much social science – even avowedly progressive forms – to appropriate without recompense the experiences of marginalised people. As Chouinard puts it:

> Contesting privileged knowledges of disabling differences require[s] research methods that go beyond giving 'voice' to the experience of persons with disabilities, to actively empowering those with disabilities in the production of geographic knowledge.
>
> (1997: 384)

It is surely this capacity to direct empowering knowledge against disabling practices and ideologies that will define an enabling Geography.

Appendix
Notes on primary sources used

The feudal case study (Chapter 5)

The problem of 'data poverty'

Historians studying the experience of the feudal peasantry must confront the general problem of data poverty. Partly, this empirical deprivation is the result of the class and gender origins and biases of chroniclers in the middle ages. Not surprisingly, the legacy of medieval record keepers is a historical account concerned mostly with the lives of the feudal patriciate (especially notable individuals). By contrast, the experiences of the feudal peasantry survive only as traces in records kept for juridical (e.g. manorial court rolls) or religious reasons. Recently, exhaustive studies of local (village and/or manorial) record sets have managed to excavate much concerning the concrete experience of peasants in certain communities (e.g., Howell, 1983; Hanawalt, 1986). None the less, the current historical picture of peasant life is a general impression, composed with broad-brush empirical strokes.

Herein lay a great difficulty for this study. While an examination of original records (such as the court rolls for a particular manor) may have been desirable for this study, such an undertaking lay well outside my scholarly capacities and interests. Feudal data sets in original form are certainly not amenable to ordinary social scientific analysis, and the elucidation of each may consume the energies of an experienced medievalist for many years. The route taken through this difficulty in the present study was first to rely upon published analyses of particular local records of feudal society. In addition, an extensive search of published primary records was undertaken in the hope that a particular data set containing some evidence of the lived experience of impairment might be located and used to substantiate the secondary source analysis. A review of medieval primary data failed to reveal records either accessible to analysis or which contained evidence of impairment.

Eventually the historical scope of the empirical search was widened to include the entire pre-industrial era, with the result that two data sets from the English County Records Series were located. These records have the dual advantages of accessibility and subject pertinence (both contain evidence of impairment). The fact that both record sets date from the early modern, rather than the feudal, era raises certain issues concerning their application to the analysis, and these are considered below. None the less, I point out that certain influential medievalists would regard both data sets as essentially feudal in character. Hilton (1985), for example, believes that the bourgeois revolution of 1640 presents an alternative terminal date for English feudalism.

Two surveys of the poor in early modern England

The two primary data collections consulted in Chapter 5 were the surveys of the poor carried out in Norwich in 1570, and in Salisbury in 1635. Strictly speaking, these surveys date from an early capitalist, rather than medieval, era. In addition, both were undertaken in towns and depart somewhat from the focus of this analysis, the rural peasant setting.

However, judicious use of these data sets can still be a powerful aid to the general analysis of the feudal social space of impairment. Several reasons may be given in support of this assertion. First, the survey periods both fall within places where capitalist social relations had gained very little purchase in the everyday lives of the subordinate classes. Only limited progress in the development of a primitive accumulative sphere is evident in the data – for example, most of the working poor identified would appear to have had control over both their labour process and their means of production (the majority are described as craft-workers working from home, no doubt with medieval technology). In addition, guild restrictions in Norwich, and perhaps in Salisbury, were still hindering the free development of capitalist trade and production.

Second, the socio-spatial constitution of both towns at the times of the surveys would seem to have preserved certain general features of the medieval landscape, especially the relative unity of work and home and the integration of poorhouses within the main living areas. The third support for the use of these data sets is a 'negative' one in that the absence of any equivalents for earlier periods heightens their importance. Comprehensive surveys of the poor, for example, were not undertaken in England until the sixteenth century.

The Norwich (1570) census is reproduced in a volume edited by Pound (1971). Pound explains that the census was undertaken as a prelude to a complete reorganisation of the city's poor law scheme in the 1570s. In the survey, 2,359 people are identified as poor, or rather less than one quarter of the population of early Elizabethan Norwich (10,625) (Pound, 1971). The census manuscript is composed of a series of descriptive entries covering the enumerated population. A sample entry is reproduced below:

> John Monde of 26 yeris, laborer and a lame man, that fyll pypes and Jone, his wyfe, of 46 yeris, that knytt, and have dwelt here 26 yere.

Poverty did not necessarily mean unemployment, with over 66 per cent of males enumerated, and more than 85 per cent of females, being in some form of employment. In addition, fewer than one quarter of the enumerated poor were deemed worthy of financial relief by the authorities. Clearly, the survey population represents a lower stratum of the settled working population, rather than a pauperised class of indigent beggars. In a more recent commentary, Pound (1988) has also noted that poverty was not synonymous with either disease or disability, with only a small minority of the listed poor being so described.

The Salisbury survey (1635) is found in a volume edited by Slack (1975). The census, undertaken by the town's justices, identified 108 households, or 249 persons, as poor. The survey is reproduced in table form, detailing the address, number of children, age, employment, weekly earnings and amount of public assistance received for the adult poor. Again the survey population cannot be said to correspond to a mendicant class. Of the enumerated poor, eighty-five (34 per cent) were recorded as earning an income (very few are people described as unemployed, and only one person is noted as begging).

The principal reason for interrogating these data sets is to probe the material situation of impairment in the pre-modern era. Both surveys were undertaken with the aim of providing civic authorities with information concerning the material circumstances of the subaltern classes in both settings. The data collected by the enumerators clearly reflect this commission, detailing the ability of those surveyed to provide successfully for their material needs. Data were collected on a range of variables – such as sex, earning capacity, possessions, age, physical impairments and affective ties – so that the poor law administrators could arrive at an implied dependency index for each enumerated household.

The industrial capitalism case study (Chapter 6)

The second set of primary materials was drawn upon in Chapter 6, focusing on the empirical setting of colonial Melbourne. In contrast to the feudal case, historical data were readily available for this spatio-temporal setting. These primary resources each represent a significant historical example of the three elements of social space – home, workplace and institution – in the colonial capitalist setting.

The Guest Biscuit Factory

The first records consulted for the colonial Melbourne case are the engagement books of the Guest and Company Biscuit, Cake and Flour Manufactory, covering the three-year period 1889–91. These data are held at the University of Melbourne Archives,Victoria, Australia. During the subject period, the firm was located in the west of central Melbourne. The manufactory was of a considerable size by colonial standards, and in 1888 employed over 100 hands (though this was reduced in 1891 with the onset of a severe local economic downturn). Most of the labour force were youths aged less than 20 years. A total of 708 new engagements were identified in the records.

Data in the engagement books include the employee's name, age, father's occupation, period of employment and home address. In addition, the records contain various annotations by the foremen which reveal something of the qualitative dimensions of the labour process. These remarks are mostly derogatory, often noting reasons for an employee's dismissal.

The Guest records are a valuable source of information about the social space of inner Melbourne in the nineteenth century. Few records as detailed as this survive from workplaces in nineteenth-century Australia, a fact which has made them an important primary data source for historians (e.g., Lee, 1988; Fox, 1991) concerned with the internal labour process of colonial factories. The data are of particular interest because they confirm the presence of the modern factory regimen in colonial Melbourne. The foremen's marginal remarks are important for their candour, which occasionally extends to the physical attributes of dismissed workers.

The Guest engagement books offer the opportunity to gauge something of the physical relationship between home and workplace in colonial Melbourne. The data indicate the locality in which workers were domiciled, providing the means for a home-workplace distance analysis. The Guest records have not been subjected to such an analysis before; indeed, with the exception of Lack's Footscray study (1980), home-to-work commuting patterns in nineteenth-century Melbourne have received very little scholarly attention. As the means for an indicator of the separation between home and work, the Guest data provide the opportunity for empirically testing the claim that workers in colonial Melbourne were forced each day to travel significant distances to the place of their employment. The test, of course, is a limited one, offering only the potential to prove that such a claim was certainly true (or false) in one instance, and, by implication, not completely false in the general case. The familiar problem of generalising from the particular is acknowledged, but, in this case, the wider accuracy of the trip indicator is partly confirmed by Lack's (1980) analysis of workers commuting to, and from, nineteenth-century Footscray.

The Melbourne Benevolent Asylum

The second data source for the colonial period is the admissions books of the city's principal nineteenth-century poorhouse, the Melbourne Benevolent Asylum (MBA). These data are held at the La Trobe Library, State Library of Victoria, Melbourne. The MBA was opened in 1851 as a refuge for the 'Aged, Infirm, Disabled and Destitute of all Creeds and Nations'.

The study period consists of the two decades, 1860–80. During this time the MBA was located in the inner suburb of North Melbourne. It was certainly not the only institution in inner Melbourne during the colonial period: by the 1880s, there were at least seventeen major philanthropic facilities in what Kennedy (1985) has called Victoria's 'charity network'. The MBA is of empirical interest because it was a colonial version of the workhouses which became an important feature in industrial England after the passage of the Poor Law Amendment Act in 1834. Workhouses tended to operate as holding receptacles for marginalised labour power, and were no doubt major features of the social spaces of many physically impaired persons.

The only other general purpose institution in colonial Melbourne was the Immigrants' Home, established in 1852. Obviously, an analysis of the populations of both the Immigrants' Home and the Benevolent Asylum would have been desirable as a means for empirically assessing patterns of institutionalisation for impaired persons in nineteenth-century Melbourne. Unfortunately, an archival search failed to reveal the existence of any Immigrants' Home records which indicated the physical condition of inmates. The admissions records from the MBA, however, were located and found to contain data on the physical status of those admitted. Importantly, these data are contained under the heading 'Ground for Application', revealing the significance accorded to impairment as a cause of social dependency.

From the data, it is possible to assess the number of impaired people who entered the MBA during the twenty-year study period. Any impairment of an applicant was recorded as a reason for admission. Consequently, one can be sure that the sub-population identified through such a discriminatory analysis was composed of persons institutionalised *because of their impairment(s)*. Obviously, this analysis cannot determine the incidence of institutionalisation among the general population of impaired people in inner Melbourne; there is no data source or investigative technique which could provide such an indicator. But the discriminatory technique can demonstrate whether or not large numbers of impaired persons experienced institutionalisation in colonial Melbourne. This analysis, when combined with various other documentary sources, including the data set introduced below, can illuminate how certain impaired poor people encountered institutions in colonial Melbourne.

Finally, a classification of impairment types may be expected to indicate common conditions (and, in certain cases, causes) associated with social dependency. Again, for these indicators the absence of knowledge concerning the overall population of impaired people in colonial Melbourne makes generalisation difficult and is thus cause for some caution. Some specific methodological issues concerning the data are now briefly discussed.

Unfortunately, the ground for application for admittance was only sporadically recorded prior to 1860, and not universally noted until 1865. It is this variable of the data which identifies whether a person was impaired. Although a substantial number of inmates had their ground of application registered after 1860, it must be assumed that the data concealed an unknown number of impaired people until universal recording began five years later. This, of course, means that the analysis of the 1860–80 admissions data will to some extent underestimate the true incidence of impairment among those received into the Asylum in that period.

Previous admissions of the same person were not necessarily recorded, and many inmates left and then returned to the Asylum several times within a single year. The task of identifying how many different individuals were admitted within any year was thus a daunting one. Consequently, I decided that a discrimination of individual inmates would only be conducted for two sample classes from the data. The first were those for whom a physical impairment was indicated. Some

597 individuals with physical impairments were admitted into the Asylum during the study period; the majority of these (77 per cent) were men.

The second class for whom a discrimination of individuals admitted was undertaken was the population of admissions taken quinquennially during the study period, beginning with the year 1860. This discrimination of all individuals admitted in a particular year was undertaken in order that a quinquennial assessment of average age at admission could be arrived at. The very difficult task of obtaining a figure for the total number of individuals admitted during the study period was beyond my resources in this study. But it is evident from the admissions records that several thousand individuals must have entered the Asylum at least once during the period 1860–80. Thus, physically impaired persons can be said to have formed a significant proportion of the total inmate population of the Asylum in the two decades following 1860.

The Melbourne Ladies' Benevolent Society

The nineteenth-century proletarian home survives only as a statistical cipher in the various public records kept by municipalities and national governments. Like their present-day successors, civic enumerators were little concerned with the inner qualities of proletarian domestic space, concentrating instead on increasing endeavours to map (and regulate) the internal and external dimensions of industrial environments. Surveillance of the domestic space of workers was a purely private enterprise, usually undertaken by voluntary legions from the middle and upper strata of the bourgeoisie. These attempts to supervise the conduct of workers' home environments were carried out under the guise of religious and secular philanthropic and educational associations that had the aim of 'improving' the lower classes. Consequently, in reviewing the domestic space of the industrial working classes, one must look to the records of these voluntary bodies, rather than those kept by government, for concrete evidence of proletarian home life.

One such set of empirical records are found in the minutes of the Melbourne Ladies' Benevolent Society (MLBS) covering the period 8 June 1850 to 19 June 1900. These data are held at La Trobe Library, State Library of Victoria, Melbourne. Established in 1845, the MLBS modelled itself on the evangelical ladies' visiting societies of Britain (Wionarski and Abbott, 1945). Dickey (1980) has described the MLBS as the largest and most efficient of colonial Australia's benevolent societies.

Importantly, the relief provided by the Society was in an 'outdoor' form (i.e., to the homes of the poor directly, rather than from an 'indoor' institutional base). The Society's five operational districts were divided into a number of sub-districts, each with a responsible 'lady visitor'. Original maps indicating the extent and location of these sub-districts appear, unfortunately, to have been lost. However, Swain (1985) has painstakingly reconstructed one supervised by a Mrs Hughes in Collingwood during the 1880s and 1890s. This indicates that sub-districts as large as one square kilometre may not have been uncommon. The visitors were each responsible for the relief of these significantly sized sub-districts, and were expected to report details of any new cases, and the progress of old ones, at the Society's fortnightly meetings.

The fine, pointed handwriting in which these minuted reports were recorded is sometimes difficult to decipher, but, happily, a complete set of manuscripts exists covering all meetings from 1850. These summary minutes provide the empirical materials for an analysis of the domestic space of the proletariat of inner Melbourne. It would appear that the Society's dealings were generally with the more settled elements of the lumpenproletariat, such as widows, families of petty criminals, prostitutes, the incapacitated and the unemployed. However, at times of general distress – such as the slump of the late 1870s and the Depression of the 1890s – elements of the more 'respectable' stratum of the working class sought help from the MLBS.

The MLBS minutes offer a rich source of empirical data on the domestic context of impairment. Obviously, the first interrogation of these records must be directed to identifying the number of impaired individuals assisted by the Society during the study period. This is necessary in order to gain some appreciation of the numbers of impaired people among the lower classes. The technique used to achieve this first enquiry was to extract and compile entries in the minutes where the visitors' re-ports indicate relief to a household with a physically impaired member. Some 1,134 entries concerning relief to households with at least one impaired member were taken from the minutes. A total of 1,004 physically impaired persons received relief from the MLBS during the study period (1850–1900).

From this analysis it is possible to classify and tabulate the types of impairments which assisted persons had. By comparing this tabulation with the corresponding data from the Benevolent Asylum, one might expect to draw conclusions about the effect of various impairments in leading to different forms of social depend-ency, experienced either in the home or in an institution. This tabulation also exposes the range of impairments which led to some level of social dependency, and thus disablement. This in turn says something of the disabling processes of mainstream work.

The data reveal many other characteristics concerning the domestic situation of impaired people, including its relationship with institutional space. The records contain references to institutions, including the MBA. Many of the people assisted by the Society spent time in institutions and the lady visitors frequently recorded their views concerning the appropriateness of institutionalising family members with physical impairments. These references in the manuscripts help to establish the external context of institutions such as the Benevolent Asylum by linking them to the domestic space of the working class.

The manuscripts also contain clues as to the relationship between home and workplace for impaired persons. For those impaired persons described as work-ing, it is possible to establish a profile of occupations and workplaces. Here two things can be shown: first, the proportion of assisted impaired people who were engaged in some form of productive activity, and second, the situation of these work activities within the wider social space of industrial Melbourne (i.e. whether marginal or mainstream).

The cultural limitations of the data are significant. The women of the MLBS confined their dealings to the white (predominantly Anglo-Celtic) proletarian communities of inner Melbourne. By 1891, the city's large Chinese community included over 1,500 people, mostly concentrated in the north-eastern quadrant of the central grid (McConville, 1985). The inner areas of the city no doubt also contained Aboriginal people who had remained after they had been dispossessed of their lands. Both communities were avoided by the Society. Requests for help from Jewish people were not ignored, but rather referred to the philanthropic associations of that community.

The Dunedin, New Zealand, case study (Chapter 9)

Primary qualitative information for this study was derived from a set of semi-structured interviews in 1995 with twenty key actors who are involved with, and affected by, the regulation of access in Dunedin. Seven of these actors were central and local government officers involved with the administration of access laws, including planners, lawyers and policy analysts. A further six interviewees came from business and community sectors, including the construction industry, legal firms and disability service organisations. In addition, seven disabled people, representing various local (including nationally affiliated) advocacy organisations were interviewed.

Due to project resource limitations, only eleven of the interviews were conducted as face-to-face meetings. These interviews were recorded with each interviewee's permission and fully transcribed. Telephone interviews were conducted with the remaining nine informants – these conversations were not formally recorded, although extensive notes were taken for each. It was decided that all of the disabled informants would be interviewed person-to-person for two reasons. First, as disabled users of the built environment, it was recognised that these informants possessed the richest experiential appreciation of the adequacy of accessibility laws in Dunedin. Second, the project had the explicit normative goal of contributing to the improvement of the accessibility laws *for disabled people*, and it was therefore critical to prioritise their views in the research. The remaining face-to-face interviews were conducted with three council building control officers and the manager of a disability services provider. In addition to the interview data, a range of secondary materials was consulted, including official reports, newspaper archives, legislation and policy documents.

Notes

1 Introduction

1 By 1998, at least three specific sessions on disability issues had been held at national geographical conferences: Association of American Geographers (AAG)
(Toronto, 1990; Chicago 1995) and Royal Geographical Society/Institute of
British Geographers (RGS-IBG) (Exeter, 1997). DAGIN was planning to spon-
sor and co-sponsor six different sessions at the 1998 meeting of the AAG. In
addition, the Geography of Health Research Group of the RGS-IBG has spon-
sored a series of disability-related conferences since 1995.
2 Volume 15, issue 4.
3 http://web.qub.ac.uk/geosci/research/geography/disbib/disgeog1.html.
4 Michael Dorn, originally based at Pennsylvania State University, and more recently at the University of Kentucky, has been the main inspirational force behind
the genesis of GEOGABLE and also, to a significant extent, DAGIN. His insightful geographic study of disability (Dorn, 1994) has also been influential, and I
shall make frequent reference to it in this book.
5 It is not necessary to quote names in these pages as I can express, and have
expressed, my gratitude privately to people whose thinking is not expressed in
ways that can be formally cited.

2 Social science and disability

1 This is to say, self-consciously organised, rather than lucid or insightful.
2 Barnes (1995: 378) has argued recently that 'most of the work on disability coming out of ... the USA ... has been bereft of theory'.
3 There are relatively few academic departments which deal exclusively with disability theory and policy in Western universities.
4 See, for example, the collection by Begum *et al.* (1994) and the recent review of
this by Oliver (1995).
5 Bickenbach's (1993) dizzying survey of conceptual approaches is a good reference work for readers wanting a more fine-grained review of disability terminology and its uses.
6 Normalisation continues to inform service policy and practice in many Western
countries: witness the volume of essays on *Normalisation in Practice* edited by
Alaszewski and Ong (1990).
7 See also Wolfensberger and Nirje (1972) for a full explanation of the principle.
8 The title of Hevey's (1992) treatise on disability, social theory and photography
suggests the abandonment of disabled people by the discipline of history.
9 These authors make the general claim that 'while modern social science devel-

oped, the disabled as a social group were ignored' (McCagg and Siegelbaum, 1989: 5).

10 At one point Oliver (1996: 35) seems to eschew any significance of the body for disability; this remark, however, is a strategic gesture, intended to distance disablement from any notion of medical causation. The twin definition of disability used by materialists instates the body as the key site where disabling social relations are registered; in short, the perspective draws upon an embodied ontology.

11 This parallels the definition adopted by Disabled People's International in 1981 (Barnes, 1991).

12 Although a seeming antonym, 'disablism' is also frequently used as a synonym for 'ableism'.

13 One important, though rather isolated, exception to this observation is the collection of essays edited by Albrecht (1981) which presented a cross-national, sociological analysis of disability policy regimes.

14 See also his recent collaborative publications (e.g., Golledge *et al.* 1996a, 1996b; Kitchin *et al.*, 1997).

15 I am indebted to the work of Dorn (1994) which alerted me to the existence of some of this research.

16 See also Moss (1997) and Dyck (1995).

17 For reasons not clear to me, the main force of Golledge's censoriousness was directed at Imrie.

3 The nature of disability

1 Rabinach (1990) also detects the implicit body in Marx's theory of nature, as a corporeal intersection of cosmos and work.

2 Kandal (1988: 91) shows that 'Marx accepted the bourgeois notion of femininity'.

3 'Nature does not produce on the one side owners of money or commodities, and on the other men possessing nothing but their own labour power. This relation has no natural basis ... It is clearly the result of a past historical development' (Marx, quoted in Smith, 1984: 48).

4 Even the early neo-classical catechist, Alfred Marshall, was moved in 1890 to describe the 'free competition' unleashed by the Industrial Revolution as 'a huge untrained monster' (1930: 11).

5 As Gallagher (1987) points out, this idea of proletarian 'weak bodies' had first been essayed by Malthus (1798), and had become something of a thematic commonplace in the writings of Victorian social commentators, such as Carlyle. Marx was the first to escape the observational vacuousness of this tradition by exposing the *social forces* which *produced* physical debility among the working class.

6 Engels (1973), of course, undertook a celebrated description of the unbearable physical pressures placed on the working class by industrial capitalism.

7 As is now common in academic texts, I use 'sic' in this instance to censure the sexist practice of reducing humanity to male pronouns. However, to signal this 'error' throughout the study in this way would confront the reader with an irritating distraction and the practice will not be proceeded with.

8 Timpanaro (1975) recognises the significance of space – one of the 'influences of the natural environment' – as a primordial material force in human life.

9 It may be added that *non-participation* in the act of labour, through class membership (e.g., the bourgeoisie of capitalism) or due to social exclusion (the 'incapable'), is also a source of social embodiment.

10 Of course, as Soper (1995) and Eagleton (1990) argue, both Nietzsche and Freud were also aware of the broader ideological significance of the mind–body dualism.

11 There is plenty of evidence to show that Foucault regarded himself, albeit loosely,

as a historical materialist *analyst* (see e.g., Fraser, 1989: 29). I am happy for the moment to bracket his rather more ambivalent support for historical materialist *politics* (cf. Harvey, 1996: 108).

12 Here I distance myself from a mundane possibilism. Burgess (1978) has criticised Possibilism for its failure to explain the social and historical creation of possibilities.

13 In the present context, this phrase must be preferable to 'labour process' in that it concedes the role of both the productive and reproductive spheres in the creation of labour power.

14 In the first sentence of her treatise on the sense of spatiality inspired by Rimbaud and the Paris Commune, Ross (1988: 3) announces her focus as 'the *social imagination* of space and time' (emphasis added).

15 This sense of 'objectivity' emerges from philosophical realism and asserts the universality of physical space as an inevitable delimiting force for human beings. All knowledge, however, of physical space is inevitably *theory-informed*.

16 There are obvious parallels between Lefebvre's idea of spatiality and Hägerstrand's 'time-geography' (Pred, 1977) insofar as both conceptions stress the role of individual practices in the creation of social space(s).

17 The 'Whig' approach to history chronicles the past as Progress: social development as a march towards the achievement of liberal democratic ideals (Livingstone, 1992).

4 Historical-geographical materialism and disability

1 See also Berkowitz (1987) and Liachowitz (1988) for alternative statist accounts which focus on the development of disability policy in the United States.

2 'The customs and code of honour of the tribe are opposed to any *individual* accumulation in excess of the average' (Mandel, 1968: 30–1) (his emphasis).

3 Liachowitz (1988) has also produced a chronicle of American disability legislation. The author alludes to a materialist position by asserting that disability is the product of the 'relationship between physically impaired individuals and their social environments' (1988: 2). However, Liachowitz later reduces this 'social environment' to its juridical content by announcing her intention to 'demonstrate how *particular laws have converted* physical deviation into social and civil disability' (1988: 3) (emphasis added). Thus, the entire material substrate of the social environment vanishes, leaving only a juridical superstructure.

4 It is timely, given this and previous criticisms, to recall here Marx's (1978: 5) warning that we cannot judge 'a period of transformation by its own consciousness; on the contrary, this consciousness must be explained rather from the contradictions of material life'.

5 The school initially comprised the group of French historians associated with the journal *Annales d'histoire économique et sociale*, established in 1929 by Lucien Febvre and Marc Bloch (Jary and Jary, 1991).

6 Or more precisely, the means of production and labour power.

7 The reference here is, of course, to Marx's oft-quoted claim that 'epochs in the history of society are no more separated from each other by strict and abstract lines than are geological epochs' (1976: 492).

8 The principal external imperatives were the tithes, inheritances, and ceremonial expenses which the peasant household had to provide for (Hanawalt, 1986).

5 The social space of disability in feudal England

1 Anderson (1974a), in fact, extends this observation to all of medieval Europe.

2 According to Pound (1988), thirteen of the families listed in the survey were actually contributing to the poor rate.

3 This figure is, of course, complicated by the Dissolution of monasteries during the 1530s; an occurrence which led to the sudden appearance of new institutions in the decades which followed, as secular hospitals were founded to replace the lost monastic centres.

6 The social space of disability in the industrial city

1 The white invasion of what is now known as the State of Victoria began in a sustained fashion in 1835. The six and a half decades which separated this date from the creation of the Commonwealth of Australia in 1901 constitute the colonial period of Victoria's history.
2 Enclosure was to prove a powerful solvent of feudal social space: by 1500, around 45 per cent of the land area of England was under private control; two centuries later, just over 70 per cent of the countryside had been enclosed (Beckett, 1990).
3 The early 1870s.
4 'Here was a vision of city growth in the classic industrial mould, departing from the pattern of Manchester and Birmingham only by its reliance on external ignition rather than the spontaneous combustion of a home-grown industrial revolution' (Davison, 1978: 6).
5 Footscray, another industrial locality within the metropolis was, by 1891, exulting in the title 'Birmingham of the South' (Davison, 1978).
6 *The Argus*, 22 October, 1863.
7 Minutes of the Melbourne Ladies' Benevolent Society (27 January 1891). Held at LaTrobe Library, State Library of Victoria, Melbourne.
8 *Report of the Chief Inspector of Factories on the 'Sweating System' in Connexion with the Clothing Trade in the Colony of Victoria*, V.P.P. 1891, vol. 3, no. 138.
9 Minutes of the Melbourne Ladies' Benevolent Society (18 December 1873).
10 Ibid. (14 January 1890).
11 Ibid. (3 November 1868).
12 Ibid. (7 December 1875).
13 Ibid. (24 March 1891).
14 Ibid. (19 May 1891).

7 Disability and the capitalist city

1 See Oliver (1991) and the collection edited by Swain *et al.* (1993) for overviews of these literatures.
2 I note that Dorn (1994: 100–104) also finds Young's schema useful in explaining disability oppression.
3 In a recent acrimonious interchange with Fraser (1997b), Young (1997) made clear that she saw the former's critique of her work as misguided rather than constructive. However, I think Young has missed the point of Fraser's criticisms, which address several ambiguities and weaknesses in her formulation of injustice.
4 Bristo (1995) reports that the Americans with Disabilities Act (1990) forced many employers to make physical accommodations in their workplaces for disabled people – 69 per cent have found that this cost nothing to do.
5 These figures have been rounded to the nearest tenth.
6 This is a very small, and therefore highly partial, sample of the recent literature on disabling imagery – other references can be found in Shakespeare's paper and also among the citations I have made in Chapter 2.
7 Butler and Bowlby (1997) suggest that the contemporary public spaces of Western cities are less tolerant of corporeal diversity than their historical equivalents in the last century. In terms of mainstream public domains, such as streets, this claim is probably true, but I think it important to keep in mind the emergent spaces of

contemporary cities, such as 'gay terrains', where resistance to corporeal conformity is regularly practised. Of course, apart from momentary occupations of spaces in protests, disabled people remain largely marginalised from the public arenas of city life.

8 This documentary, 'Vital Signs: Crip Culture Talks Back', was made by David Mitchell and Sharon Snyder, English Department, Northern Michigan University, Marquette, MI 49855, USA.

9 There is a voluminous literature which both supports these assertions, and highlights the pervasiveness of these discriminations in Western cities generally. This literature cannot be surveyed in entirety here; however, useful starting sources are Swain *et al.* (1993) (UK), Minister for Health, Housing and Community Services (1991) (Australia) and Eastern Bay of Plenty People First Committee (1993) (New Zealand). Lunt and Thornton (1994) also provide an authoritative overview of employment and disability in fifteen Western countries.

10 My information comes from ADAPT press releases posted on the GEOGABLE listserv on 15 November 1997.

11 Dorn (1994) also correctly observes that the United States disability movement lacks the critical social theoretical appreciation of its British counterpart.

8 Community care: the environment of justice?

1 My analysis of community care is essentially an empirical investigation framed by one political-ethical principle, enabling justice. For a fuller, ethical analysis of the ideals of community care, see Wilmot's (1997) important book.

2 A year later, there were further alarming allegations of systematic mistreatment in New South Wales institutions, including an allegation that disabled residents had on occasion been chained up and fed dog food (*Sydney Morning Herald*, 27 November 1997: 3).

3 This was known as the 'Better Cities Program'.

4 This information was related by Marsha Coleman of ADAPT and posted on the GEOGABLE listserv on 15 November 1997.

5 My information comes from ADAPT press releases posted on the GEOGABLE listserv on 15 November 1997.

6 Since 1994, the institution had been 'auspiced by a charitable organisation, Oberlin Ltd., although the for profit company (Kanowana Pty. Ltd.) that previously ran it is currently contracted to provide administrative services and is also the landlord' (Community Services Commission, 1997: 1).

7 Many disability support programmes in New Zealand are funded through the Ministry of Health.

9 The regulation of urban accessibility

1 This research was funded by a grant from the University of Otago. Two research assistants, Megan Turnbull and Leah McBey, collected and transcribed the data for the case study – their invaluable help is gratefully acknowledged. The helpful co-operation of the study informants is also acknowledged.

2 Fry's rather isolated earlier study of disabling cartographic practices provided an important precedent for the investigations of mapping and disability by Matthews and Vujakovic (1995).

3 The Race Relations Act 1971 and the Human Rights Commission Act 1977.

4 It is not unlawful to treat disabled people differently if the intention is to ensure that their 'special needs' are met.

5 Another important area of exclusion is the government's immunity from compliance with the HRA. The state is not bound to uphold the legislation in the same

way as other members of the community – a fact which has caused considerable concern in the broader community.
6 Letter of DCC Building Control Manager to Otago University Disabilities Co-ordinator, Ms. Donna-Rose Mackay, dated 23 December 1993. Copy supplied by Ms. Mackay.
7 Which in any case had been made clear in previous press statements – see earlier discussion.
8 Personal communication with Donna-Rose Mackay, past (1995) President of Dunedin Branch of Disabled Persons' Assembly, 14 May 1996.
9 Compliance Schedule Audit undertaken by New Zealand Fire Service (Rotorua District) for South Waikato District Council, September/October 1995. Copy of Audit avaliable from South Waikato District Council.
10 To be fair to Dorn, he does recognise that the enactment of the ADA has not eliminated all forms of disability discrimination (e.g. 1994: 209–11).

10 Towards an enabling geography

1 Of course, the separate discriminations arising from gender and disability overlap within individual social identities, causing what some feminists have called a 'double handicap' for disabled women (see Lonsdale, 1990).
2 This literature is simply too voluminous now to be cited here exhaustively. Good overviews of the work on institutional care by geographers are to be found in Dear and Wolch (1987), Philo (1998) and the collection edited by Smith and Giggs (1988).
3 The work of Wolpert (e.g., 1978) is an exception to this observation.
4 An 'externality' is the spillover effect from any land use activity. 'Distance decay' refers to the tendency of externalities to decline in intensity with increasing distance from their point of origin.
5 Evidence for this statement is admittedly anecdotal, but I think none the less persuasive, and includes the rising participation of non-geographers in the listserv, GEOGABLE. One can also cite the increasing co-operation between geographers and non-geographers in a range of conferences and seminars on disability issues, especially in the United States.

References

Abberley, P. (1985) 'Policing Cripples: Social Theory and Physical Handicap', unpublished paper, copy obtained from author.

Abberley, P. (1987) 'The Concept of Oppression and the Development of a Social Theory of Disability', *Disability, Handicap, and Society*, 2, 1, 5–19.

Abberley, P. (1991a) *Disabled People: Three Theories of Disability*. Occasional Papers in Sociology, no. 10, Bristol: Department of Economics and Social Science, Bristol Polytechnic.

Abberley, P. (1991b) *Handicapped by Numbers: A Critique of the OPCS Disability Surveys*. Occasional Papers in Sociology, no. 9, Bristol: Department of Economics and Social Science, Bristol Polytechnic.

Abberley, P. (1993) 'Disabled People and "Normality" ' in Swain, J., Finkelstein, V., French, S. and Oliver, M. (eds), *Disabling Barriers – Enabling Environments*, London: Sage.

Abberley, P. (1997) 'The Spectre at the Feast – Disabled People and Social Theory', unpublished paper, copy obtained from author.

Abbott, M.W. and Kemp, D.R. (1993) 'New Zealand', in Kemp, D.R., (ed.), *International Handbook on Mental Health Policy*, Westport, Connecticut: Greenwood, 217–251.

Adas, M. (1989) *Machines as the Measure of Men: Science, Technology, and Ideologies of Western Dominance*, Ithaca: Cornell University Press.

Ainley, R. (ed.) (1998) *New Frontiers of Space, Bodies and Gender*, London: Routledge.

Alaszewski, A. and Ong, B.N. (1990) *Normalisation in Practice*, London: Tavistock/Routledge.

Albrecht, G.L. (ed.) (1981) *Cross National Rehabilitation Policies: a Sociological Perspective*, London: Sage.

Alcock, P. (1993) *Understanding Poverty*, London: Macmillan.

Allen, I. (1992) 'Purchasing and Providing: What Kind of Progress?' in Allen, I., (ed.), *Purchasing and Providing Social Services in the 1990s: Drawing the Line*, London: Policy Studies Institute, 1–6.

Anderson, E. (1979) *The Disabled Schoolchild*, London: Methuen.

Anderson, P. (1974a) *Passages from Antiquity to Feudalism*, London: New Left Books.

Anderson, P. (1974b) *Lineages of the Absolutist State*, London: New Left Books.

Appleby, Y. (1994) 'Out in the Margins', *Disability and Society*, 9, 1, 19–32.

Arbeitsgemeinschaft der Deutschen Hauptfürsorgestellen (1995) *Das ABC der Behindertenhilfe*, Köln: ACON.

Archer, T. (1985 [1865]) *The Pauper, the Thief and the Convict*, New York: Garland.

Ashton, T.S. (1948) *The Industrial Revolution, 1760–1830*, London: Oxford University Press.

Ault, W.O. (1965) 'Open-Field Husbandry and the Village Community: A Study of

Agrarian By-Laws in Medieval England', *Transactions of the American Philosophical Society*, 55, 7.

Ault, W.O. (1972) *Open-Field Farming in Medieval England: A Study of Village By-Laws*, London: Allen & Unwin.

Australian Bureau of Statistics (1993) *Disability, Ageing and Carers Australia, 1993: Summary of Findings*, Canberra: Australian Government Publishing Service.

Australian Bureau of Statistics (1995) *Focus on Families: Caring in Families – Support for Persons Who Are Older or Have Disabilities*, Canberra: Australian Government Publishing Service.

Bachelard, G. (1969) *The Poetics of Space*, Boston: Beacon Press.

Badcock, B. (1984) *Unfairly Structured Cities*, Oxford: Basil Blackwell.

Bain, S.M. (1971) 'The Geographical Distribution of Psychiatric Disorders in the North East Region of Scotland', *Geographia Medica: International Journal of Medical Geography*, 2, 84–108.

Baldwin, S. (1993) *The Myth of Community Care: an Alternative Neighbourhood Model of Care*, London: Chapman and Hall.

Balukas, R. and Baken, J.W. (1985) 'Community Resistance to Development of Group Homes for People with Mental Retardation', *Rehabilitation Literature*, 46, 7–8, 194–197.

Barker, M. (1981) 'Human Biology and the Possibility of Socialism', in Mepham, J. and Ruben, D-H. (eds), *Issues in Marxist Philosophy – Volume Four: Social and Political Philosophy*, Brighton: Harvester Press.

Barnes, C. (1991) *Disabled People in Britain and Discrimination: a Case for Anti-Discrimination Legislation*, London: Hurst and Co.

Barnes, C. (1992a) 'Disability and Employment', *Personnel Review*, 21, 6, 55–73.

Barnes, C. (1992b) *Disabling Imagery and the Media*, Halifax: Ryburn/BCODP.

Barnes, C. (1995) 'Review of "Disability is Not Measles"', *Disability and Society*, 10, 3, 378–81.

Barnes, C. (1996) 'Foreword', in Campbell, J. and Oliver, M., *Disability Politics: Understanding Our Past, Changing Our Future*, London: Routledge.

Barrett, B. (1971) *The Inner Suburbs: The Evolution of an Industrial Area*, Melbourne: Melbourne University Press.

Barretta-Herman, A. (1994) *Welfare State to Welfare Society: Restructuring New Zealand's Social Services*, New York: Garland.

Beamish, C. (1981) 'State, Space and Crisis: Towards a Theory of the Public City in North America', unpublished M.A. thesis, McMaster University.

Bean, P. (1988) 'Mental Health in Europe: Some Recent Trends', in Smith, C.J. and Giggs, J.A. (eds), *Location and Stigma: Contemporary Perspectives on Mental Health and Mental Health Care*, Boston: Unwin Hyman.

Beckett, J.V. (1990) *The Agricultural Revolution*, Oxford: Blackwell.

Begum, N., Hill, M. and Stevens, A. (eds) (1994) *Reflections: Views of Black Disabled People on their Lives and Community Care*, London: CCETSW.

Beier, A.L. (1983) *The Problem of the Poor in Tudor and Early Stuart England*, London: Methuen.

Beier, A.L. (1985) *Masterless Men: The Vagrancy Problem in England, 1560–1640*, London: Methuen.

Benjamin, G.J. (1981) 'Group Homes and Single-Family Zoning', *Zoning and Planning Law Report*, 4, 97–102.

Bennett, T. (1990) 'Planning and People with Disabilities', in Montgomery, J. and Thornley, A. (eds), *Radical Planning Initiatives: New Directions for Planning in the 1990s*, Aldershot: Gower.

Bennie, G. (1993) 'Deinstitutionalisation: Critical Factors for Successful Transition to the Community', report produced for Central Regional Health Authority, Wellington, New Zealand.

Benthall, J. (1976) *The Body Electric: Patterns of Western Industrial Culture*, London: Thames and Hudson.

Berdiansky, H.A. and Parker, R. (1977) 'Establishing a Group Home for the Adult Mentally Retarded in North Carolina', *Mental Retardation*, 15, 8–11.

Berg, M. (1988) 'Women's Work, Mechanization and Early Industrialization', in Pahl, R.E. (ed.), *On Work: Historical, Comparative and Theoretical Approaches*, Oxford: Blackwell.

Berkowitz, E.D. (1987) *Disabled Policy: America's Programs for the Handicapped*, Cambridge: Cambridge University Press.

Berkowitz, M. and Hill, M.A. (1989) 'Disability and the Labor Market: an Overview', in Berkowitz, M. and Hill, M.A. (eds), *Disability and the Labor Market*, Ithaca, NY: ILR Press.

Berthoud, R., Lakey, J. and McKay, S. (1993) *The Economic Problems of Disabled People*, London: Policy Studies Institute.

Bewley, C. and Glendinning, C. (1994) 'Representing the Views of Disabled People in Community Care Planning', *Disability and Society*, 9, 3, 301–314.

Bickenbach, J. (1993) *Physical Disability and Social Policy*, Toronto: University of Toronto Press.

Blank, R.H. (1994) *New Zealand Health Policy: a Comparative Study*, Oxford: Oxford University Press.

Blaut, J.M. (1976) 'Where Was Capitalism Born?', *Antipode*, 8, 2, 1–11.

Bloch, M. (1962) *Feudal Society*, 2 vols., London: Routledge & Kegan Paul.

Bloch, M. (1967) *Land and Work in Mediaeval Europe*, London: Routledge & Kegan Paul.

Boston, J. (1992) 'Redesigning New Zealand's Welfare State', in Boston, J. and Dalziel, P. (eds), *The Decent Society? Essays in Response to National, Social and Economic Policies*, Auckland: Oxford University Press.

Bottomore, T., Harris, L., Kiernan, V.G. and Milliband, R. (eds) (1983), *A Dictionary of Marxist Thought*, Oxford: Blackwell.

Bovi, A. (1971) *Breugel*, London: Thames and Hudson.

Boylan, E. (ed) (1991) *Women and Disability*, London: Zed.

Brail, R., Hughes, J. and Arthur, C. (1976) *Transportation Services for the Disabled and Elderly*, New Brunswick, NJ: Center for Urban Policy and Research.

Braudel, F. (1973) *Capitalism and Material Life, 1400–1800*, London: Weidenfeld & Nicolson.

Braudel, F. (1981) *Civilization and Capitalism – Volume One: 15th–18th Century. The Structures of Everyday Life: The Limits of the Possible*, London: Collins.

Briggs, A. (1959) *The Age of Improvement, 1783–1867*, London: Longman.

Briggs, A. (1968) *Victorian Cities*, Harmondsworth: Penguin.

Bristo, M. (1995) 'Lessons from the Americans with Disabilities Act', in Zarb, G. (ed.), *Removing Disabling Barriers*, London: Policy Studies Institute.

Brooke, C.N.L. (1978) 'Both Small and Great Beasts: An Introductory Study', in Baker, D. (ed.), *Medieval Women*, Oxford: Blackwell.

Brown, K. (1996) 'Caring for People with Severe and Multiple Disabilities', in Department of Human Services and Health (ed.), *Towards a National Agenda for Carers: Workshop Papers*, Canberra: Australian Government Publishing Service.

Brown-May, A. (1995) 'The Highway of Civilisation and Common Sense: Street Regulation and the Transformation of Social Space in 19th and Early 20th Century Melbourne', Working Paper No. 49, Urban Research Program, Australian National University.

Browning, D. (1992) 'Purchaser/Provider Split: Passing Fashion or Permanent Fixture?', in Allen, I., (ed.), *Purchasing and Providing Social Services in the 1990s: Drawing the Line*, London: Policy Studies Institute.

Buck, N.H. (1981) 'The Analysis of the State Intervention in Nineteenth-Century

Cities: The Case of Municipal Labour Policy in East London' in Dear, M. and Scott, A.J. (eds), *Urbanization and Urban Planning in Capitalist Society*, London: Methuen.

Bundesministerium für Arbeit und Sozialordnung (1997) *Ratgeber für Behinderte*, Bonn.

Burgess, R. (1978) 'The Concept of Nature in Geography and Marxism', *Antipode*, 10, 2, 1–11.

Burkhauser, R.V. (1989) 'Disability Policy in the United States, Sweden and the Netherlands', in Berkowitz, M. and Hill, M.A. (eds), *Disability and the Labor Market*, Ithaca, NY: ILR Press.

Burnett, A. and Moon, G. (1983) 'Community Opposition to Hostels for Single Homeless Men', *Area*, 15, 161–166.

Butler, R.E. (1994) 'Geography and Vision Impaired and Blind Populations', *Transactions, Institute of British Geographers*, 19, 366–368.

Butler, R.E. and Bowlby, S. (1997) 'Bodies and Spaces: an Exploration of Disabled People's Experiences of Public Space', *Environment and Planning D: Society and Space*, 15, 4, 411–433.

Buttimer, A. (1969) 'Social Space in Interdisciplinary Perspective', *Geographical Review*, 59, 417–426.

Cahill, M. (1991) *Exploring the Experience of Disability*, Wellington: Ministry of Health.

Campbell, J. and Oliver, M. (1996) *Disability Politics: Understanding Our Past, Changing Our Future*, London: Routledge.

Campbell, S. (1994) 'The Valued Norm: Supported Accommodation for People with Disabilities: A Discussion Paper', Sydney: New South Wales. Department of Community Services, Ageing and Disability Services Directorate.

Campling, J. (1981) *Images of Ourselves*, London: Routledge & Kegan Paul.

Cass, B., Gibson, F. and Tito, F. (1988) *Social Security Review – Towards Enabling Policies: Income Support for People with Disabilities*, Canberra: Australian Government Publishing Service.

Checkland, S.G. and Checkland, E.O.A. (1974) 'Introduction', in Checkland, S.G. and Checkland, E.O.A. (eds), *The Poor Law Report of 1834*, Harmondsworth: Penguin.

Chouinard, V. (1994) 'Reinventing Radical Geography: Is All That's Left Right?' *Environment and Planning D: Society and Space*, 12, 1, 2–6.

Chouinard, V. (1997) 'Making Space for Disabling Differences: Challenging Ableist Geographies', *Environment and Planning D: Society and Space*, 15, 4, 379–387.

Chouinard, V. and Grant, A. (1995) 'On Being Not Even Anywhere Near "The Project": Ways of Putting Ourselves in the Picture', *Antipode*, 27, 2, 137–166.

Clapham, D. and Kintrea, K. (1992) *Housing Co-Operatives in Britain: Achievements and Prospects*, Harlow: Longman.

Clapham, D., Kemp, P. and Smith, S.J. (1990) *Housing and Social Policy*, London: Macmillan.

Clay, R.M. (1909) *The Mediaeval Hospitals of England*, London: Methuen.

Collier, A. (1979) 'Materialism and Explanation in the Human Sciences', in Mepham, J. and Ruben, D-H. (eds), *Issues in Marxist Philosophy – Volume Two: Materialism*, Brighton: Harvester.

Community Services Commission (1997) *Suffer the Children: the Hall for Children Report*, Sydney: Community Services Commission.

Consulting Group (1992) *Impact of Willow Croft on Residential House Values in Waverley*, unpublished report, available from University of Otago Commercial Consulting Group, PO Box 56, Dunedin, New Zealand.

Cook, I. (1991) 'Drowning in See-World? Critical Ethnographies of Blindness', unpublished MA Thesis, University of Kentucky, Lexington, KY.

Cooper, M. (1990) *Women and Disability*, Canberra: Disabled People's International.

Corker, M. (1993) 'Integration and Deaf People: the Policy and Power of Enabling Environments', in Swain, J., Finkelstein, V., French, S. and Oliver, M. (eds), *Disabling Barriers – Enabling Environments*, London: Sage.

Cormode, L. (1997) 'Emerging Geographies of Disability and Impairment: an Introduction', *Environment and Planning D: Society and Space*, 15, 4, 387–390.

Cullingworth, J.B. (1985) *Town and Country Planning*, 9th ed., London: Allen & Unwin.

Currie, R.F., Trute, B., Tefft, B. and Segall, A. (1989) 'Maybe on My Street: the Politics of Community Placement of the Mentally Disabled', *Urban Affairs Quarterly*, 25, 2, 298–321.

Dalley, G. (ed.) (1991) *Disability and Social Policy*, London: Policy Studies Institute.

Daniel, C. (1998) 'Radical, Angry and Willing to Work', *New Statesman*, 6 March, 22–23.

Davis, K. (1996) 'Disability and Legislation: Rights and Equality', in Hales, G. (ed.), *Beyond Disability: Toward an Enabling Society*, London: Sage.

Davis, L.J. (1995) *Enforcing Normalcy: Disability, Deafness, and the Body*, London: Verso.

Davis, L.J. (ed.) (1997) *The Disability Studies Reader*, New York: Routledge.

Davison, G. (1978) *The Rise and Fall of Marvellous Melbourne*, Melbourne: Melbourne University Press.

Dear, M. (1977) 'Spatial Externalities and Locational Conflict', in Massey, D.B. and Batey, P.W.J. (eds), *Alternative frameworks for analysis*, London Papers in Regional Science 7, London: Pion.

Dear, M. (1980) 'The Public City', in Clark, W.A.V. and Moon, E.G. (eds), *Residential Mobility and Public Policy*, Beverly Hills: Sage.

Dear, M. (1981) 'Social and Spatial Reproduction of the Mentally Ill', in Dear, M. and Scott, A.J. (eds), *Urbanization and Urban Planning in Capitalist Societies*, New York: Methuen, 481–497.

Dear, M. (1992) 'Understanding and Overcoming the NIMBY Syndrome', *Journal of the American Planning Association*, 58, 3, 288–299.

Dear, M. and Gleeson, B.J. (1991) 'Attitudes Towards Homelessness: the Los Angeles Experience', *Urban Geography*, 12, 2, 155–176.

Dear, M. and Laws, G. (1986) 'Anatomy of a Decision: Recent Land Use Zoning Appeals and Their Effect on Group Home Locations in Ontario', *Canadian Journal of Community Mental Health*, 5, 1, 5–17.

Dear, M. and Taylor, S.M. (1982) *Not On Our Street: Community Attitudes to Mental Health Care*, London: Pion.

Dear, M. and Wolch, J. (1987) *Landscapes of Despair: From Deinstitutionalization to Homelessness*, Cambridge: Polity.

Dear, M, Fincher, R. and Currie, L. (1977) 'Measuring the External Effects of Public Programs', *Environment and Planning A*, 9, 137–147.

Dear, M., Gaber, S.L., Takahashi, L. and Wilton, R. (1997) 'Seeing People Differently: the Sociospatial Construction of Disability', *Environment and Planning D: Society and Space*, 15, 4, 455–480.

Dear, M., Taylor, S.M. and Hall, G.B. (1980) 'External Effects of Mental Health Facilities', *Annals of the Association of American Geographers*, 70, 3, 342–352.

Dear, M., Wolch, J. and Wilton, R. (1994) 'The Human Service Hub Concept in Human Services Planning', *Progress in Planning*, 42, 3, 174–271.

Deegan, M.J. and Brooks, N.A. (eds) (1985) *Women and Disability: The Double Handicap*, New Brunswick: Transaction Books.

DeHoog, R.H. (1984) *Contracting Out for Human Services: Economic, Political and Organizational Perspectives*, Albany, NY: State University of New York Press.

Demone, H.W. and Gibelman, M. (1989) 'In Search of a Theoretical Base for the Purchase of Services', in Demone, H.W. and Gibelman, M. (eds), *Services for Sale: Purchasing Health and Human Services*, New Brunswick, NJ: Rutgers University Press.

de Neufville, J.I. (1981) 'Land Use: a Tool for Social Policies', in de Neufville, J.I., (ed.), *The Land Use Policy Debate in the United States*, New York: Plenum.

Dettwyler, K.A. (1991) 'Can Paleopathology Provide Evidence for "Compassion"?', *American Journal of Physical Anthropology*, 84, 375–384.

Dickey, B. (1980) *No Charity There: A Short History of Social Welfare in Australia*, Melbourne: Thomas Nelson.

Disability Alliance (1987a) *Disability Rights Bulletin*, London: Disability Alliance.

Disability Alliance (1987b) *Poverty and Disability: Breaking the Link*, London: Disability Alliance.

Dodds, A.G. (1980) 'Spatial Representation and Blindness', unpublished Ph.D. thesis, University of Nottingham.

Doray, B. (1988) *From Taylorism to Fordism: A Rational Madness*, London: Free Association.

Dorn, M. (1994) 'Disability as Spatial Dissidence: A Cultural Geography of the Stigmatized Body', unpublished M.Sc. thesis, The Pennsylvania State University.

Doyal, L. (1993) 'Human Need and the Moral Right to Optimal Community Care', in Bornat, J., Pereira, C., Pilgrim, D. and Williams, F. (eds), *Community Care: a Reader*, London: Macmillan.

Doyal, L. and Gough, I. (1991) *A Theory of Human Need*, London: Macmillan.

Drake, R.F. (1997) 'What am I Doing here? "Non-Disabled" People and the Disability Movement', *Disability and Society*, 12, 4, 643–645.

Driver, F. (1993) *Power and Pauperism: the Workhouse System, 1834–1884*, Cambridge: Cambridge University Press.

Duby, G. (1968) *Rural Economy and Country Life in the Medieval West*, Columbia, SC: University of South Carolina Press.

Duncan, N. (1996) *BodySpace: Destabilising Geographies of Gender and Sexuality*, London: Routledge.

Durkheim, E. (1964) *The Division of Labour in Society*, New York: The Free Press.

Dyck, I. (1995) 'Hidden Geographies: the Changing Lifeworlds of Women with Disabilities', *Social Science and Medicine*, 40, 307–320.

Eagleton, T. (1988) 'Foreword', in Ross, K. *The Emergence of Social Space: Rimbaud and the Paris Commune*, London: Macmillan.

Eagleton, T. (1990) *The Ideology of the Aesthetic*, Oxford: Blackwell.

Eastern Bay of Plenty People First Committee (1993) 'People First Conference Report 1993', Copy available from People First, PO Box 3017, Ohope Eastern Bay of Plenty, New Zealand.

Edwards, M.L. (1997) 'Deaf and Dumb in Ancient Greece', in Davis, L.J. (ed.) *The Disability Studies Reader*, New York: Routledge.

Eisenstein, Z.R. (ed.) (1979) *Capitalist Patriarchy and the Case for Socialist Feminism*, New York: Monthly Review Press.

Elliget, T. (1988) 'The New South Wales Richmond Programme of Deinstitutionalization and the Voluntary Sector', *Community Mental Health in New Zealand*, 4, 1, 13–20.

Engels, F. (1973) *The Condition of the Working-Class in England from Personal Observation and Authentic Sources*, Moscow: Progress Publishers.

England, K. (1994) 'Getting Personal: Reflexivity and Positionality in Feminist Research', *The Professional Geographer*, 46, 1, 80–89.

Evans-Pritchard, E. (1937) *Witchcraft, Oracles and Magic amongst the Azande*, Oxford: Clarendon Press.

Eyles, J. (1988) 'Mental Health Services, the Restructuring of Care, and the Fiscal Crisis of the State: the United Kingdom Case Study', in Smith, C.J. and Giggs, J.A. (eds), *Location and Stigma: Contemporary Perspectives on Mental Health and Mental Health Care*, Boston: Unwin Hyman.

Fincher, R. (1978) 'Some Thoughts on Deinstitutionalization and Difference', *Antipode*, 10, 1, 46–50.

Fincher, R. (1991) 'Caring for Workers' Dependants: Gender, Class and Local State Practice in Melbourne', *Political Geography Quarterly*, 10, 4, 356–381.

Fine, M. and Asch, A. (eds) (1988) *Women with Disabilities: Essays in Psychology, Culture and Politics*, Philadelphia: Temple University Press.

Finger, A. (1991) *Past Due: a Story of Disability, Pregnancy and Birth*, London: The Women's Press.

Finger, A. (1995) '"Welfare Reform" and Us', *Ragged Edge*, November/December, 15 and 36.

Finkelstein, V. (1980) *Attitudes and Disabled People*, New York: World Rehabilitation Fund.

Finkelstein, V. and Stuart, O. (1996) 'Developing New Services', in Hales, G. (ed.), *Beyond Disability: Toward an Enabling Society*, London: Sage.

Foldvary, F. (1994) *Public Goods and Private Communitites: the Market Provision of Public Services*, Aldershot: Edward Elgar.

Foote, T. (1971) *The World of Breugel, c.1525–1569*, New York: Time-Life International.

Ford, S. (1996) 'Learning Difficulties', in Hales, G. (ed.), *Beyond Disability: Toward an Enabling Society*, London: Sage.

Foreman, P.J. and Andrews, G. (1988) 'Community Reaction to Group Homes', *Interaction*, 2, 5, 15–18.

Forester, J. (1989) *Planning in the Face of Power*, Berkeley: University of California Press.

Foucault, M. (1975) *The Birth of the Clinic: An Archaeology of Medical Perception*, New York: Vintage.

Foucault, M. (1979) *Discipline and Punish: The Birth of the Prison*, New York: Vintage.

Foucault, M. (1980a) *Power/Knowledge: Selected Interviews and Other Writings, 1972–1977*, New York: Pantheon.

Foucault, M. (1980b) *The History of Sexuality (vol.1): An Introduction*, New York: Vintage.

Foucault, M. (1986) *The History of Sexuality (vol.2): The Use of Pleasure*, New York: Vintage.

Foucault, M. (1988a) *The History of Sexuality (vol.3): The Care of the Self*, New York: Vintage.

Foucault, M. (1988b) *Madness and Civilisation: A History of Insanity in the Age of Reason*, New York: Vintage.

Fougere, G. (1994) 'Health', in Spoonley, P., Pearson, D. and Shirley, I. (eds), *New Zealand Society: a Sociological Introduction*, 2nd edn, Palmerston North: Dunmore.

Fox, C. (1991) *Working Australia*, Sydney: Allen & Unwin.

Frank, A.W. (1990) 'Bringing Bodies Back In: A Decade Review' *Theory, Culture and Society*, 17, 1, 131–162.

Fraser, N. (1989) *Unruly Practices: Power, Discourse and Gender in Contemporary Social Theory*, Polity.

Fraser, N. (1995) 'From Redistribution to Recognition? Dilemmas of Justice in a "Post-Socialist" Age', *New Left Review*, 212, 68–73.

Fraser, N. (1997a) *Justice Interruptus: Reflections on the 'Postsocialist' Condition*, New York: Routledge.

Fraser, N. (1997b) 'A Rejoinder to Iris Young', *New Left Review*, 223, May/June, 126–130.

Freeman, J. (1888) *Lights and Shadows of Melbourne Life*, London: Sampson Low, Marston Searle & Rivington.

French, S. (1993a) 'Disability, Impairment or Something in Between?', in Swain, J., Finkelstein, V., French, S. and Oliver, M. (eds) *Disabling Barriers – Enabling Environments*, London: Sage.

French, S. (1993b) 'What's so Great about Independence?', in Swain, J., Finkelstein, V., French, S. and Oliver, M. (eds) *Disabling Barriers – Enabling Environments*, London: Sage.

Freudenberg, N. (1984) *Not in Our Backyards! Community Action for Health and the Environment*, New York: Monthly Review Press.

Gallagher, C. (1987) 'The Body Versus the Social Body in the Works of Thomas Malthus and Henry Mayhew', in Gallagher, C. and Laqueur, T. (eds), *The Making of the Modern Body: Sexuality and Society in the Nineteenth Century*, Berkeley: University of California Press.

Gallagher, C. and Laqueur, T. (1987) 'Introduction', in Gallagher, C. and Laqueur, T. (eds), *The Making of the Modern Body: Sexuality and Society in the Nineteenth Century*, Berkeley: University of California Press.

Gant, R. (1992) 'Transport for the Disabled', *Geography*, 77, 1, 88–91.

Gant, R. and Smith, J. (1984) 'Spatial Mobility Problems of the Elderly and Disabled in the Cotswolds', in Clark, G. *et al.* (eds), *The Changing Countryside. Proceedings of the First British–Dutch Symposium on Rural Geography*, Norwich: Geo Books.

Gant, R.L. and Smith, J.A. (1988) 'Journey Patterns of the Elderly and Disabled in the Cotswolds: a Spatial Analysis', *Social Science and Medicine*, 27, 2, 173–180.

Gant, R.L. and Smith, J.A. (1990) 'Feet First in Kingston Town Centre: A Study of Personal Mobility', Kingston Accessibility Studies Working Paper no. 2, Kingston Polytechnic (UK).

Gant, R. and Smith, J. (1991) 'The Elderly and Disabled in Rural Areas: Travel Patterns in the North Cotswolds', in Champion, T. and Wadkins, C. (eds) *People in the Countryside*, London: Paul Chapman, 108–124.

Garland, R. (1995) *The Eye of the Beholder: Deformity and Disability in the Graeco-Roman World*, London: Duckworth.

Gartner, A. and Joe, T. (eds) (1987) *Images of the Disabled, Disabling Images*, New York: Praeger.

Genicot, L. (1966) 'Crisis: From the Middle Ages to Modern Times', in Postan, M.M. (ed.), *The Cambridge Economic History of Europe – Volume One: The Agrarian Life of the Middle Ages*, Cambridge: Cambridge University Press.

George, M. (1995) 'Broken Promises', *Community Care*, 24–30 August, 1082, 16.

Gething, L. (1997) 'Sources of Double Disadvantage for People with Disabilities Living in Remote and Rural Areas of New South Wales, Australia', *Disability and Society*, 12, 4, 513–531.

Gies, F. and Gies, J. (1990) *Life in a Medieval Village*, New York: Harper & Row.

Giggs, J.A. (1973) 'The Distribution of Schizophrenics in Nottingham', *Transactions, Institute of British Geographers*, 59, 55–76.

Gilderbloom, J.I. and Rosentraub, M.S. (1990) 'Creating the Accessible City: Proposals for Providing Housing and Transportation for Low Income, Elderly and Disabled People', *American Journal of Economics and Sociology*, 49, 3, 271–282.

Gilligan, C. (1982) *In a Different Voice: Psychological Theory and Women's Development*, Cambridge, MA: Harvard University Press.

Gleeson, B.J. (1993) 'Second Nature? The Socio-spatial Production of Disability', unpublished Ph.D thesis, The University of Melbourne.

Gleeson, B.J. (1995a) 'Disability – a State of Mind?', *Australian Journal of Social Issues*, 29, 1, 10–23.

Gleeson, B.J. (1995b) 'A Space for Women: the Case of Charity in Colonial Melbourne', *Area*, 27, 3, 193–207.

Gleeson, B.J. (1996a) 'A Geography for Disabled People?', *Transactions, Institute of British Geographers*, 21, 2, 387–396.

Gleeson, B.J. (1996b) 'Disability Studies – a Historical Materialist View', *Disability and Society*, 12, 2, 179–202.

Gleeson, B.J. (1996c) 'Let's Get Planning out of Community Care', *Urban Policy and Research*, 14, 3, 227–229.

Gleeson, B.J. (1997) 'The Regulation of Environmental Accessibility in New Zealand', *International Planning Studies*, 2, 3, 367–390.

Gleeson, B.J. (1998) 'Disability and Poverty', in Fincher, R. and Nieuwenhuysen, J. (eds), *Australian Poverty: Then and Now*, Melbourne: Melbourne University Press.

Gleeson, B.J. and Memon, P.A. (1994) 'The NIMBY Syndrome and Community Care Facilities: a Research Agenda for Planning', *Planning Practice and Research*, 9, 2, 105–118.

Gleeson, B.J. and Memon, P.A. (1997) 'Community Care: Implications for Urban Planning from the New Zealand Experience', *Planning Practice and Research*, 12, 2, 119–32.

Gleeson, B.J., Gooder, H.F. and Memon, P.A. (1995) *Community Care Facilities: a Guide for Planners and Service Providers*, Dunedin: Environmental Policy and Management Research Centre, University of Otago.

Glendinning, C. (1991) 'Losing Ground: Social Policy and Disabled People in Great Britain, 1980–90', *Disability, Handicap and Society*, 6,1, 3–19.

Godelier, M. (1978) 'System, Structure and Contradiction in "Capital"', in McQuarie, D. (ed.), *Marx: Sociology/Social Change/Capitalism*, London: Quartet.

Goffman, E. (1964) *Stigma, Notes on the Management of Identity*, Harmondsworth: Penguin.

Goffman, E. (1969) *Strategic Interaction*, Philadelphia: University of Pennsylvania Press.

Golledge, R.G. (1990) 'Special Populations in Contemporary Urban Regions', in J. F. Hart (ed.) *Our Changing Cities*, Baltimore: Johns Hopkins University Press.

Golledge, R.G. (1991) 'Tactual Strip Maps as Navigational Aids', *Journal of Visual Impairment and Blindness*, 85, 7, 296–301.

Golledge, R.G. (1993) 'Geography and the Disabled: a Survey with Special Reference to Vision Impaired and Blind Populations' *Transactions, Institute of British Geographers*, 18,1, 63–85.

Golledge, R.G. (1996) 'A Response to Gleeson and Imrie', *Transactions, Institute of British Geographers*, 21, 2, 404–411.

Golledge, R.G. (1997) 'On Reassembling One's Life: Overcoming Disability in the Academic Environment', *Environment and Planning D: Society and Space*, 15, 4, 391–409.

Golledge, R.G., Klatzky, R.L. and Loomis, J.M. (1996a) 'Cognitive Mapping and Wayfinding by Adults without Vision', in J. Portugali (ed.), *The Construction of Cognitive Maps*, Dordrecht: Kluwer.

Golledge, R.G., Costanzo, C.M. and Marston, J.R. (1996b) 'Public Transit Use By Non-Driving Disabled Persons: The Case of the Blind and Visually Impaired', California Path Working Paper.

Golledge, R.G., Loomis, J.M., Klatzky, R.L., Flury, A. and Yang, X-L. (1991) 'Designing a Personal Guidance System to Aid Navigation without Sight: Progress on the GIS Component', *International Journal of Geographical Information Systems*, 5, 373–396.

Goodall, B. (1987) *Dictionary of Human Geography*, London: Penguin.

Gordon, E.E. (1983) 'Epithets and Attitudes', *Archives of Physical Medicine Rehabilitation*, 64, 234–235.

Gottdiener, M. (1985) *The Social Production of Urban Space*, Austin: University of Texas Press.

Gregory, D. (1981) 'Human Agency and Human Geography', *Transactions, Institute of British Geographers*, 6, 1, 1–16.

Grob, G.N. (1995) 'The Paradox of Deinstitutionalization', *Society*, July/August, 51–59.

Grosz, E.A. (1992) 'Bodies-Cities', in Colomina, B. (ed.), *Sexuality and Space*, New York: Princeton Architectural Press.

Hahn, H. (1986) 'Disability and the Urban Environment: a Perspective on Los Angeles', *Environment and Planning D: Society and Space*, 4, 273–288.

Hahn, H. (1987a) 'Accepting the Acceptably Employable Image: Disability and Capitalism', *Policy Studies Journal*, 15, 3, 551–570.

Hahn, H. (1987b) 'Civil Rights for Disabled Americans: the Foundation of a Political Agenda', in Gartner, A. and Joe, T. (eds), *Images of the Disabled/Disabling Images*, New York: Praeger.

Hahn, H. (1988) 'Can Disability Be Beautiful?', *Social Policy*, Winter, 26–32.

Hahn, H. (1989) 'Disability and the Reproduction of Bodily Images: The Dynamics of Human Appearances', in Wolch, J. and Dear, M. (eds) *The Power of Geography: How Territory Shapes Social Life*, Boston: Unwin Hyman.

Haj, F. (1970) *Disability in Antiquity*, New York: Philosophical Library.

Hales, G., (ed.) (1996) *Beyond Disability: Toward an Enabling Society*, London: Sage.

Hall, E.C. (1994) *Researching Disability in Geography*, Spatial Policy Analysis Working Paper 28, Manchester: School of Geography, University of Manchester.

Hall, E. (1997) 'Work Spaces: Refiguring the Disability–Employment Debate', Exeter: Royal Geographical Society/Institute of British Geographers Annual Conference.

Hanawalt, B. (1986) *The Ties That Bound: Peasant Families in Medieval England*, New York: Oxford University Press.

Hanks, J.R. and Hanks, L.M. (1948) 'The Physically Handicapped in Certain Non-Occidental Societies', *The Journal of Social Issues*, 4, 4, 11–20.

Harding, S. (1992) 'After the Neutrality Ideal: Politics, Science and "Strong Objectivity"', *Social Research*, 59, 3, 567–587.

Harris, D-R., Rowlands, M., Ballard, K., Smith, K. and Gleeson, B.J. (1995) 'Disability and Tertiary Education in New Zealand', unpublished report submitted to Ministry of Education (New Zealand).

Harrison, J. (1987) *Severe Physical Disability: Responses to the Challenge of Care*, London, Cassell.

Harrison, M. and Gilbert, S. (eds) (1992) *The Americans with Disabilities Handbook*, Beverly Hills, CA: Excellent Books.

Hartmann, H. (1979) 'Capitalism, Patriarchy, and Job Segregation by Sex', in Eisenstein, Z.R. (ed.), *Capitalist Patriarchy and the Case for Socialist Feminism*, New York: Monthly Review Press.

Harvey, D. (1982) *The Limits to Capital*, Oxford: Blackwell.

Harvey, D. (1989a) *The Urban Experience*, Oxford: Blackwell.

Harvey, D. (1989b) 'From Managerialism to Entrepreneurialism: the Transformation in Urban Governance in Late Capitalism', *Geografiska Annaler*, 71B, 1, 3–17.

Harvey, D. (1990) 'Between Space and Time: Reflections on the Geographical Imagination', *Annals of the Association of American Geographers*, 80, 3, 418–434.

Harvey, D. (1993) 'Class Relations, Social Justice, and the Politics of Difference', in Keith, M. and Pile, S. (eds), *Place and the Politics of Identity*, London: Routledge.

Harvey, D. (1996) *Justice, Nature and the Politics of Difference*, Oxford: Blackwell.

Hayek, F. (1979) *Social Justice, Socialism and Democracy: Three Australian Lectures*, Sydney: Centre for Independent Studies.

Hearn, K. (1991) 'Disabled Lesbians and Gays Are Here to Stay', in Kaufman, T. and Lincoln, P. (eds), *High Risk Lives: Lesbian and Gay Politics after the Clause*, Bridport: Prism Press.

Heginbotham, C. (1990) *The Return to Community: the Voluntary Ethic and Community Care*, London: Bedford Square.

Hekman, S. (1995) *Moral Voices, Moral Selves: Carol Gilligan and Feminist Moral Theory*, University Park, PA: Pennsylvania State University Press.

Helvarg, D. (1995) 'Legal Assault on the Environment', *The Nation*, 30 January, 126–127.

Herlihy, D. (1968) *Medieval Culture and Society*, New York: Harper & Row.

Hevey, D. (1992) *The Creatures Time Forgot: Photography and Disability Imagery*, London: Routledge.

Hevey, D. (1997) 'The Enfreakment of Photography', in Davis, L.J. (ed.) *The Disability Studies Reader*, New York: Routledge.

Higgins, W. (1982) 'To Him that Hath …: The Welfare State', in Kennedy, R. (ed.), *Australian Welfare History: Critical Essays*, Melbourne: Macmillan.

Hill, M.H. (1985) 'Bound to the Environment: Towards a Phenomenology of Sightlessness', in Seamon, D. and Mugerauer, R. (eds) *Dwelling Place and Environment: Towards a Phenomenology of Person and World*, New York: Columbia University Press.

Hill, M.H. (1986) *The Nonvisual Lifeworld: a Comparative Phenomenology of Blindness*, unpublished Ph.D. thesis, Kent, OH: Kent State University.

Hillyer, B. (1993) *Feminism and Disability*, Norman: University of Oklahoma Press.

Hilton, R.H. (1975) *The English Peasantry in the Later Middle Ages. The Ford Lectures for 1973 and Related Studies*, Oxford: Clarendon.

Hilton, R.H. (1985) 'A Crisis of Feudalism', in Aston, T.H. and Philpin, C.H.E. (eds), *The Brenner Debate: Agrarian Class Structure and Economic Development in Pre-Industrial Europe*, Cambridge University Press, Cambridge.

HMSO (1990) *NHS and Community Care Act*, London: HMSO.

Hobsbawm, E.J. (1968) *Industry and Empire: An Economic History of Britain since 1750*, London: Weidenfeld & Nicolson.

Hobsbawm, E.J. (1984) *Workers: Worlds of Labor*, New York: Pantheon.

Holden, L. (1991) *Forms of Disability*, Sheffield: JSOT Press.

Horner, A. (1994) 'Leaving the Institution', in Ballard, K. (ed.), *Disability, Family, Whanau and Society*, Hamilton North: Dunmore.

Horton, S. (1996) 'The Dunedin Mayoral Election: A Symbol of Uneven Development', *New Zealand Geographer*, 53, 1, 30–40.

Howell, C. (1983) *Land, Family and Inheritance in Transition: Kibworth Harcourt, 1280–1700*, Cambridge: Cambridge University Press.

Hoyes, L. and Means, R. (1993) 'Markets, Contracts and Social Care Services: Prospects and Problems', in Bornat, J., Pereira, C., Pilgrim, D. and Williams, F. (eds), *Community Care: a Reader*, London: Macmillan.

Hugg, L. (1979) 'A Map Comparison of Work Disability and Poverty in the United States', *Social Science and Medicine*, 13D, 237–240.

Hurst, R. (1995) 'International Perspectives and Solutions', in Zarb, G. (ed.), *Removing Disabling Barriers*, London: Policy Studies Institute.

Ignatieff, M. (1978) *A Just Measure of Pain: The Penitentiary in the Industrial Revolution, 1750–1850*, New York: Columbia University Press.

Illich, I. (1986) 'Body History', *The Lancet*, ii, 1325–1327.

Imrie, R.F. (1996a) *Disability and the City: International Perspectives*, London: Paul Chapman.

Imrie, R.F. (1996b) 'Equity, Social Justice, and Planning for Access and Disabled People: an International Perspective', *International Planning Studies*, 1, 1, 17–34.

Imrie, R.F. (1996c) 'Ableist Geographers, Disablist Spaces: Towards a Reconstruction of Golledge's "Geography and the Disabled"', *Transactions, Institute of British Geographers*, 21, 2, 397–403.

Imrie, R.F. and Wells, P.E. (1993a) 'Disablism, Planning and the Built Environment', *Environment and Planning C: Government and Policy*, 11, 2, 213–231

Imrie, R.F. and Wells, P.E. (1993b) 'Creating Barrier-Free Environments', *Town and Country Planning*, 61,10, 278–281.

Ingstad, B. and Whyte, S.R. (eds) (1995) *Disability and Culture*, Berkeley: University of California Press.

Iveson, K. (1997) 'Review of Ruddick, S. *Young and Homeless in Hollywood: Mapping Social Identities*', copy obtained from author, Urban Research Program, Australian National University, Canberra, ACT 0200, Australia.

Iveson, K. (1998) 'Putting the Public Back into the Public Sphere', *Urban Policy and Research*, 16, 1, 21–33.

Jaffe, M. and Smith, T.P. (1986) *Siting Homes for Developmentally Disabled Persons*, Chicago: American Planning Association.

Jary, D. and Jary, J. (1991) *Dictionary of Sociology*, London: Harper-Collins.

Jenkins, R. (1991) 'Disability and Social Stratification', *British Journal of Sociology*, 42, 4, 557–580.

John, A.V. (1986) 'Introduction', in John, A.V. (ed.), *Unequal Opportunities: Women's Employment in England, 1800–1918*, Oxford: Blackwell.

Johnson, L.C. (1989a) 'Embodying Geography – Some Implications of Considering the Sexed Body in Space', *Proceedings of the 15th New Zealand Geography Conference*, Dunedin: University of Otago.

Johnson, L.C. (1989b) 'Weaving Workplaces: Sex, Race and Ethnicity in the Australian Textile Industry', *Environment and Planning A*, 21, 681–684.

Johnston, R.J. (1993) 'The Rise and Decline of the Corporate-Welfare State: a Comparative Analysis in Global Context', in Taylor, P.J., (ed.), *Political Geography of the Twentieth Century: a Global Analysis*, London: Belhaven.

Johnston, R.J., Gregory, D. and Smith, D. (eds) (1994) *The Dictionary of Human Geography*, 3rd edn, Oxford: Blackwell.

Joseph, A.E. and Hall, G.B. (1985) 'The Locational Concentration of Group Homes in Toronto', *The Professional Geographer*, 37, 2, 143–154.

Joseph Rowntree Foundation (1997) *Foundations: Making Housing and Community Care Work* (Research Bulletin), York: Joseph Rowntree Foundation.

Kandal, T.R. (1988) *The Woman Question in Classical Sociological Theory*, Miami: Florida International University Press.

Kavka, G.S. (1992) 'Disability and the Right to Work', *Social Philosophy and Policy*, 9, 1, 262–290.

Kearns, D. (1983) 'A Theory of Justice – and Love: Rawls on the Family', *Politics*, 18, 2, 36–42.

Kearns, R. (1990) *Coping and Community Life for People with Chronic Mental Disability in Auckland*, Occasional Paper no. 26, Auckland: Department of Geography, University of Auckland.

Kearns, R., Smith, C.J. and Abbott, M.W. (1991) 'Another Day in Paradise? Life on the Margins in Urban New Zealand', *Social Science and Medicine*, 33, 4, 369–379.

Kearns, R., Smith, C.J. and Abbott, M.W. (1992) 'The Stress of Incipient Homelessness', *Housing Studies*, 7, 4, 280–298.

Kelsey, J. (1995) *The New Zealand Experiment*, Auckland: Auckland University Press.

Kemp, D.R. (ed.) (1993), *International Handbook on Mental Health Policy*, Westport, CT: Greenwood.

Kennedy, R. (1982) 'Introduction: Against Welfare', in Kennedy, R. (ed.), *Australian Welfare History: Critical Essays*, Melbourne, Macmillan.

Kennedy, R. (1985) *Charity Warfare: The Charity Organization Society in Colonial Melbourne*, Melbourne: Hyland.

Kiernan, M.J. (1983) 'Ideology, Politics, and Planning: Reflections on the Theory and Practice of Urban Planning', *Environment and Planning B: Planning and Design*, 10, 71–87.

Kindred, M., Cohen, J., Penrod, D. and Shaffer, T. (eds) (1976) *The Mentally Retarded Citizen and the Law*, New York: Free Press.

Kirby, A., Bowlby, S.R. and Swann, V. (1983) 'Mobility Problems of the Disabled', *Cities*, 3, 117–119.

Kitchin, R. (1997) 'Participatory Action Research in Geography: Towards a More Emancipatory and Empowering Approach', unpublished paper, copy obtained from author.

Kitchin, R. (1998) 'Out of Place', 'Knowing One's Place: Space, Power and the Exclusion of Disabled People', *Disability and Society*, forthcoming.

Kitchin, R.M., Blades, M. and Golledge, R. (1997) 'Understanding Spatial Concepts at the Geographic Scale without the Use of Vision', *Progress in Human Geography*, 21, 2, 225–242.

Korda, M. and Neumann, P. (eds) (1997) *Stadtplanung für Menschen mit Behinderungen*, Arbeitsgemeinschaft Angewandte Geographie Münster e.V., Arbeitsberichte 28.

Kosminsky, E.A. (1956) *Studies in the Agrarian History of England in the Thirteenth Century*, Oxford: Blackwell.

Kramer, R. (1986) 'The Future of Voluntary Organizations in Social Welfare', in Independent Sector Inc., *Philanthropy, Voluntary Action, and the Public Good*, Washington DC: Independent Sector Inc.

Kridler, C. and Stewart, R.K. (1992a) 'Access for the Disabled 1', *Progressive Architecture*, 73, 7, 41–42.

Kridler, C. and Stewart, R.K. (1992b) 'Access for the Disabled 2', *Progressive Architecture*, 73, 8, 35–36.

Kridler, C. and Stewart, R.K. (1992c) 'Access for the Disabled 3', *Progressive Architecture*, 73, 9, 45–46.

Kristeva, J. (1982) *The Powers of Horror: an Essay on Abjection*, New York: Columbia University Press.

Kumar, K. (1988) 'From Work to Employment and Unemployment', in Pahl, R.E. (ed.), *On Work: Historical, Comparative and Theoretical Approaches*, Oxford: Blackwell.

Labarge, M.W. (1986) *Women in Medieval Life: A Small Sound of the Trumpet*, London: Hamish Hamilton.

Lack, J. (1980) 'Residence, Workplace, Community: Local History in Metropolitan Melbourne', *Historical Studies*, 19, 74.

Lack, J. (1991) *A History of Footscray*, Melbourne: Hargreen/City of Footscray.

Lakin, K.C. and Bruininks, R.H. (eds) (1985) *Strategies for Achieving Community Integration of Developmentally Disabled Persons*, Baltimore: P.H. Brookes.

Langenfelt, G. (1954) *The Historic Origin of the Eight Hours Day*, Stockholm: Almquist and Wiksell.

Laslett, P. (1971) *The World We Have Lost*, London: Methuen.

Laura, R.S. (ed.) (1980) *The Problem of Handicap*, Melbourne: Macmillan.

Law, R.M. and Wolch, J.R. (1993) 'Homelessness and the Cities: Local Government Policies and Practices in Southern California', Los Angeles Homelessness Project, Working Paper 44, Department of Geography, University of Southern California.

Lawrence, D. (1993) 'Being Without Seeing', unpublished research project, Department of Geography, University of Waikato.

Laws, G. (1993) '"The Land of Old Age": Society's Changing Attitudes towards Urban Built Environments for Elderly People', *Annals of the Association of American Geographers*, 83, 4, 672–693.

Laws, G. (1994) 'Oppression, Knowledge and the Built Environment', *Political Geography*, 13, 1, 7–32.

Laws, G. and Dear, M. (1988) 'Coping in the Community: a Review of Factors Influencing the Lives of Deinstitutionalised Ex-Psychiatric Patients', in Smith,

C.J. and Giggs, J.A. (eds), *Location and Stigma: Contemporary Perspectives on Mental Health and Mental Health Care*, Boston: Unwin Hyman.

Laws, G. and Lord, S. (1990) 'The Politics of Homelessness', in Kodras, J.E. and Jones, J.P. (eds), *Geographic Dimensions of United States Social Policy*, New York: Edward Arnold.

Lazonick, W. (1990) *Competitive Advantage on the Shop Floor*, Cambridge, MA: Harvard University Press.

Leat, D. (1995) *The Development of Community Care by the Independent Sector*, London: Policy Studies Institute.

Lebovich, W.L. (1993) *Design for Dignity: Studies in Accessibility*, New York: Wiley.

Le Breton, J. (1985) *Residential Services and People with a Disability*, Canberra: Australian Government Publishing Service.

Leccese, M. (1993) 'Is Access Attainable?', *Landscape Architecture*, 83, 6, 71–75.

Lee, J. (1988) 'The Marks of Want and Care', in Burgmann, V. and Lee, J. (eds), *Making a Life: A People's History of Australia since 1788*, Melbourne: McPhee Gribble.

Lefebvre, H. (1979) 'Space: Social Product and Use Value', in Freiburg, J.W. (ed.), *Critical Sociology: European Perspectives*, New York: Irvington.

Lefebvre, H. (1991) *The Production of Space*, Oxford: Blackwell.

Le Goff, J. (1988) *Medieval Civilisation, 400–1500*, Oxford: Blackwell.

Le Heron, R. and Pawson, E. (eds) (1996) *Changing Places: New Zealand in the Nineties*, Auckland: Longman Paul.

Leonard, E.M. (1965) *The Early History of English Poor Relief*, London: Frank Cass.

Leonard, P. (1984) *Personality and Ideology: Towards a Materalist Understanding of the Individual* , London: Macmillan.

Lerman, P. (1981) *Deinstitutionalization and the Welfare State*, New Brunswick, NJ: Rutgers University Press.

Lewis, J. and Glennerster, H. (1996) *Implementing the New Community Care*, Buckingham: Open University Press.

Liachowitz, C.H. (1988) *Disability as Social Construct: Legislative Roots*, Philadelphia: University of Pennsylvania Press.

Lifchez, R. (ed.) (1987) *Rethinking Architecture: Design Students and Physically Disabled People*, Berkeley, CA: University of California Press.

Lifchez, R. and Winslow, B. (1979) *Design for Independent Living: the Environment and Physically Disabled People*, London: Architectural Press.

Livingstone, D. (1992) *The Geographical Tradition: Episodes in the History of a Contested Enterprise*, Oxford: Blackwell.

Loader, B. and Burrows, R. (1994) 'Towards a Post-Fordist Welfare State? The Restructuring of Britain, Social Policy and the Future of Welfare' in Burrows, R. and Loader, B. (eds), *Towards a Post-Fordist Welfare State?*, London: Routledge.

Locker, D., Rao, B. and Weddell, J.M. (1979) 'The Community Reaction to a Hostel for the Mentally Handicapped', *Social Science and Medicine*, 13A, 817–821.

Logan, J.R. and Molotch, H.L. (1987) *Urban Fortunes: the Political Economy of Place*, Berkeley: University of California Press.

Lonsdale, S. (1990) *Women and Disability: The Experience of Physical Disability among Women*, London: Macmillan.

Lovett, A.A. and Gatrell, A.C. (1988) 'The Geography of Spina Bifida in England and Wales', *Transactions, Institute of British Geographers*, 13, 288–302.

Lucas, A.M. (1983) *Women in the Middle Ages: Religion, Marriage and Letters*, Brighton: Harvester.

Lunt, N. and Thornton, P. (1994) 'Disability and Employment: Towards an Understanding of Discourse and Employment', *Disability and Society*, 9(2), 223–238.

Lynn, P. (1990) *Administrators and Change in the Penal System in Victoria, 1850–80*, unpublished Ph.D. thesis, Deakin University.

Lyons, M. (1995) 'The Development of Quasi-Vouchers in Australia's Community Services', *Policy and Politics*, 23, 2, 127–139.

McCagg, W.O. and Siegelbaum, L. (1989) *The Disabled in the Soviet Union: Past and Present, Theory and Practice*, Pittsburgh: University of Pittsburgh Press.

McCalman, J. (1984) *Struggletown: Public and Private Life in Richmond, 1900–1965*, Melbourne: Melbourne University Press.

McConnell, S. (1981) *Theories for Planning: an Introduction*, London: Heinemann.

McConville, C. (1985) 'Chinatown', in Davison, G., Dunstan, D. and McConville, C. (eds), *The Outcasts of Melbourne: Essays in Social History*, Sydney: Allen & Unwin.

McGovern, P. (1989) 'Protecting the Promise of Community-Based Care', in Demone, H.W. and Gibelman, M. (eds), *Services for Sale: Purchasing Health and Human Services*, New Brunswick, NJ: Rutgers University Press.

McIntosh, M.K. (1991) 'Treatment of the Poor in Late Medieval England', unpublished paper, University of Colorado, Boulder.

Mackenzie, S. and Rose, D. (1983) 'Industrial Change, the Domestic Economy and Home Life', in Anderson, J., Duncan, S. and Hudson, R. (eds), *Redundant Spaces? Social Change and Industrial Decline in Cities and Regions*, London: Academic Press.

Macfarlane, A. (1996) 'Aspects of Intervention: Consultation, Care, Help and Support', in Hales, G. (ed.) *Beyond Disability: Towards an Enabling Society*, London: Sage.

McLoughlin, J.B. (1994) 'Centre or Periphery? Town Planning and Spatial Political Economy', *Environment and Planning A*, 26, .1111–1122

McQuarie, D. (1978) 'Introduction', in McQuarie, D. (ed.), *Marx: Sociology/Social Change/Capitalism*, London: Quartet.

McTavish, F. (1992) 'The Effectiveness of People with Disabilities in the Policy Process: The Total Mobility Scheme Example', unpublished MA thesis, University of Otago.

Mairs, N. (1995) 'On Being a Cripple', in Petersen, L.H., Brereton, J.C. and Hartman, J.E. (eds), *The Norton Reader*, New York: Norton.

Malcomson, R.W. (1988) 'Ways of Getting a Living in Eighteenth-Century England', in Pahl, R.E. (ed.), *On Work: Historical, Comparative and Theoretical Approaches*, Oxford: Blackwell.

Malin, N. (1987) 'Community Care: Principles, Policy and Practice', in Malin, N. (ed.), *Reassessing Community Care*, London: Croom Helm.

Mandel, E. (1968) *Marxist Economic Theory*, London: Merlin.

Mangen, S.P. (1985) *Mental Health Care in the European Community*, London: Dover.

Marshall, A. (1930) *Principles of Economics: An Introductory Volume*, London: Macmillan.

Martin, L. and Gaster, L. (1993) 'Community Care Planning in Wolverhampton', in Smith, R., Gaster, L., Harrison, L., Martin, L., Means, R. and Thistlethwaite, P. (eds) *Working Together for Better Community Care*, Bristol: School of Advanced Urban Studies.

Martins, R.M. (1982) 'The Theory of Social Space in the Work of Henri Lefebvre', in Forrest, R., Henderson, J. and Williams, P. (eds), *Urban Political Economy and Social Theory: Critical Essays in Urban Studies*, London: Gower.

Marx, K. (1973) *Grundrisse*, London: Penguin.

Marx, K. (1975) *The Poverty of Philosophy*, Moscow: Progress Publishers.

Marx, K. (1976) *Capital: A Critique of Political Economy – Volume One*, London: Penguin.

Marx, K. (1977) *Economic and Philosophic Manuscripts of 1844*, Moscow: Progress Publishers.

Marx, K. (1978) 'Preface to "A Contribution to the Critique of Political Economy"', in Tucker, R.C. (ed.), *The Marx-Engels Reader*, 2nd edn, New York: Norton.

Marx, K. (1981) *Capital: A Critique of Political Economy – Volume Three*, London: Penguin.

Marx, K. and Engels, F. (1967) *The Communist Manifesto*, London: Penguin.

Marx, K. and Engels, F. (1976) *The German Ideology*, Moscow: Progress Publishers.

Marx, K. and Engels, F. (1979) *Pre-Capitalist Socio-Economic Formations: A Collection*, Moscow: Progress Publishers.

Massey, D. (1984) 'Introduction: Geography Matters', in Massey, D. and Allen, J. (eds) *Geography Matters! A Reader*, Cambridge: Cambridge University Press.

Matthews, M.H. and Vujakovic, P. (1995) 'Private Worlds and Public Places: Mapping the Environmental Values of Wheelchair Users', *Environment and Planning A*, 27, 1069–1083.

Mayer, D. (1981) 'Geographical Clues about Multiple Sclerosis', *Annals of the Association of American Geographers*, 71, 1, 28–39.

Mayhew, T. (1968a) *London Labour and the London Poor*, vol. 1, New York: Dover.

Mayhew, T. (1968b) *London Labour and the London Poor*, vol. 2, New York: Dover.

Mayhew, T. (1968c) *London Labour and the London Poor*, vol. 4, New York: Dover.

Meekosha, H. (1989) 'Research and the State: Dilemmas of Feminist Practice', *Australian Journal of Social Issues*, 24, 4, 249–268.

Memon, P.A. and Gleeson, B.J. (1995) 'Towards a New Planning Paradigm? Reflections on New Zealand's New Resource Management Act', *Environment and Planning B: Planning and Design*, 22, 109–124.

Mendus, S. (1993) 'Different Voices, Still Lives: Problems in the Ethics of Care', *Journal of Applied Philosophy*, 10, 1, 17–27.

Merleau-Ponty, M. (1962) *Phenomenology of Perception*, London: Routledge & Kegan Paul.

Meyerson, L. (1988) 'The Social Psychology of Physical Disability: 1948 and 1988', *Journal of Social Issues*, 44, 1, 173–188.

Middleton, C. (1988) 'The Familiar Fate of the *Famulae*: Gender Divisions in the History of Wage Labour', in Pahl, R.E. (ed.), *On Work: Historical, Comparative and Theoretical Approaches*, Oxford: Blackwell.

Miller, D., Rowlands, M. and Tilley, C. (eds) (1997) *Domination and Resistance: One World, One Archaeology, Volume 3*, London: Routledge.

Milligan, C. (1996) 'Service Dependent Ghetto Formation – a Transferable Concept?', *Health and Place*, 2, 4, 199–211.

Milner, A. (1993) *Cultural Materialism*, Melbourne: Melbourne University Press.

Minister for Health, Housing and Community Services (Australia) (1991) *Social Justice for People with Disabilities*, Canberra: Australian Government Publishing Service.

Mitchell, D.T. and Snyder, S.L. (eds) (1997) *The Body and Physical Difference: Discourses of Disability*, Ann Arbor: University of Michigan Press.

Moon, G. (1988) '"Is There One Around Here?" – Investigating Reaction to Small Scale Mental Health Hostel Provision in Portsmouth, England', in Smith, C.J. and Giggs, J.A. (eds), *Location and Stigma: Contemporary Perspectives on Mental Health and Mental Health Care*, Boston: Unwin Hyman.

Morris, J. (1989) *Able Lives – Women's Experience of Paralysis*, London: The Women's Press.

Morris, J. (1991) *Pride against Prejudice: Transforming Attitudes to Disability*, London: The Women's Press.

Morris, J. (1992) 'Personal and Political: a Feminist Perspective on Researching Physical Disability', *Disability, Handicap and Society*, 7, 2, 157–166.

Morris, J. (1993a) *Independent Lives? Community Care and Disabled People*, Basingstoke: Macmillan.

Morris, J. (1993b) '"Us" and "Them"': Feminist Research and Community Care', in Bornat, J., Pereira, C., Pilgrim, D. and Williams, F. (eds), *Community Care: a Reader*, London: Macmillan.

Morris, J. (ed.) (1996) *Encounters with Strangers: Feminism and Disability*, London: The Women's Press.

Morris, P. (1969) *Put Away*, London: Routledge & Kegan Paul.

Morrison, E. and Finkelstein, V. (1993) 'Broken Arts and Cultural Repair: the Role of Culture in the Empowerment of Disabled People', in Swain, J., Finkelstein, V., French, S. and Oliver, M. (eds) *Disabling Barriers – Enabling Environments*, London: Sage.

Moss, P. (1997) 'Negotiating Spaces in Home Environments: Older Women Living with Arthritis', *Social Science and Medicine*, 45, 1, 23–33.

Moss, P. and Dyck, I. (1996) 'Inquiry into Environment and Body: Women, Work, and Chronic Illness', *Environment and Planning D: Society and Space*, 14, 737–753.

Mowrey, M. and Redmond T. (1993) *Not in Our Backyard: the People and Events that Shaped America's Modern Environmental Movement*, New York: Morrow.

Mumford, L. (1961) *The City in History*, London: Pelican.

Napolitano, S. (1996) 'Mobility Impairment', in Hales, G. (ed.), *Beyond Disability: Towards an Enabling Society*, London: Sage.

Nast, H. and Pile, S. (eds) (1998) *Places Through the Body*, London: Routledge.

National Capital Authority (1996) *Institutional Reform*, Occasional Paper Series 2, Paper 5, Canberra: National Capital Authority.

Neale, R.S. (1975) 'Introduction', in Kamenka, E. and Neale, R.S. (eds), *Feudalism, Capitalism and Beyond*, Canberra: ANU Press.

Nelson, J.L. and Berens, B.S. (1997) 'Spoken Daggers, Deaf Ears and Silent Mouths: Fantasies of Deafness in Early Modern England', in Davis, L.J. (ed.) *The Disability Studies Reader*, New York: Routledge.

Nicholson, G. (1988) 'The Village in History', in Nicholson, G. and Fawcett, J. (eds), *The Village in History*, London: Weidenfeld & Nicholson.

Norden, M.F. (1994) *The Cinema of Isolation: a History of Physical Disability in the Movies*, New Brunswick, NJ: Rutgers University Press.

Nutley, S.D. (1980) 'Accessibility, Mobility and Transport-Related Welfare: The Case of Rural Wales', *Geofrum*, 11, 335–352.

Nutley, S.D. (1990) *Unconventional Community Transport in the United Kingdom*, New York: Gordon & Breach.

Oliver, M. (1986) 'Social Policy and Disability: Some Theoretical Issues', *Disability, Handicap and Society*, 1, 1, 5–17.

Oliver, M. (1990) *The Politics of Disablement*, London: Macmillan.

Oliver, M. (1991) 'Disability and Participation in the Labour Market', in Brown, P. and Scase, R. (eds), *Poor Work: Disadvantage and the Division of Labour*, Milton Keynes: Open University Press.

Oliver, M. (1992) 'Changing the Social Relations of Research Production', *Disability, Handicap, and Society*, 7, 2, 101–114.

Oliver, M. (1993) 'Disability and Dependency: a Creation of Industrial Societies?', in Swain, J., Finkelstein, V., French, S. and Oliver, M. (eds), *Disabling Barriers – Enabling Environments* London: Sage.

Oliver, M. (1995) 'Review of "Reflections …"', *Disability and Society*, 10, 3, 369–371.

Oliver, M. (1996) *Understanding Disability: From Theory to Practice*, London: Macmillan.

Oliver, M. and Barnes, C. (1993) 'Discrimination, Disability and Welfare: From Needs to Rights', in Swain, J., Finkelstein, V., French, S. and Oliver, M. (eds), *Disabling Barriers – Enabling Environments*, London: Sage.

Olson, J.M. and Brewer, C.A. (1997) 'An Evaluation of Color Selections to Accommodate Map Users with Color-Vision Impairments', *Annals of the Association of American Geographers* 87,1, 103–134.

Orr, K. (1984) 'Consulting Women with Disabilities', *Australian Disability Review*, 3, 14–18.

Pahl, R.E. (1988) 'Editor's Introduction: Historical Aspects of Work, Employment, Unemployment and the Sexual Division of Labour', in Pahl, R.E. (ed.), *On Work: Historical, Comparative and Theoretical Approaches*, Oxford: Blackwell.

Park, D. (1995) 'An Imprisoned Text: Reading the Canadian Mental-Handicap Asylum', unpublished Ph.D. thesis, York University.

Park, D.C. and Radford, J. (1997) 'Space, Place and the Asylum: an Introduction', *Health and Place*, 3, 2, 71–72.

Park, D.C., Radford, J.P. and Vickers, M.H. (1998) 'Disability Studies in Human Geography', *Progress in Human Geography*, 22, 2, 208–233.

Parker, G. (1993) 'A Four-Way Stretch? The Politics of Disability and Caring', in Swain, J., Finkelstein, V., French, S. and Oliver, M. (eds), *Disabling Barriers – Enabling Environments*, London: Sage, 249–256.

Parr, H. (1997a) 'Naming Names: Brief Thoughts on Geography and Disability', *Area*, 29, 2, 173–176.

Parr, H. (1997b) 'Mental Health, Public Space, and the City: Questions of Individual and Collective Access', *Environment and Planning D: Society and Space*, 15, 4, 35–454.

Pateman, C. (1980) ' "The Disorder of Women": Women, Love, and the Sense of Justice', *Ethics*, 19, 1, 20–34.

Pati, G.C. and Stubblefield, G. (1990) 'The Disabled are Able to Work', *Personnel Journal*, December, 30–34.

Pawson, E. (1996) 'Policy, Local Governance and the Regions', in Le Heron, R. and Pawson, E. (eds), *Changing Places: New Zealand in the Nineties*, Auckland: Longman.

Pennington, S. and Westover, B. (1989) *A Hidden Workforce: Homeworkers in England, 1850–1985*, London: Macmillan.

Perle, E. D. (1969) *Urban Mobility Needs of the Handicapped: an Exploration*, unpublished Ph.D. thesis, University of Pittsburgh.

Perske, R. and Perske, M. (1980) *New Life in the Neighbourhood: How Persons with Retardation Can Help Make a Good Community Better*, Nashville TN: Abingdon.

Philo, C. (1997) 'The "Chaotic Spaces" of Medieval Madness: Thoughts on the English and Welsh Experience' in Teich, M., Porter, R. and Gustafsson, B. (eds) *Nature and Society in Historical Context*, Cambridge: Cambridge University Press.

Philo, C. (1998) 'Across the Water: Reviewing Geographical Studies of Asylums and Other Mental Health Facilities', *Health and Place*, 3, 2, 73–89.

Pile, S. (1996) *The Body and the City: Psychoanalysis, Space and Subjectivity*, London: Routledge.

Pile, S. and Keith, M. (eds) (1997) *Geographies of Resistance*, London: Routledge.

Pinch, S. (1985) *Cities and Services: the Geography of Collective Consumption*, London: Routledge.

Pinch, S. (1997) *Worlds of Welfare*, London: Routledge.

Plotkin, S. (1987) *Keep Out: the Struggle for Land Use Control*, Berkeley, CA: University of California Press.

Pollard, S. (1963) 'Factory Discipline in the Industrial Revolution', *Economic History Review*, 2nd series, 16, 254–271.

Postan, M.M. (1966) 'England', in Postan, M.M. (ed.), *The Cambridge Economic History of Europe – Volume One: The Agrarian Life of the Middle Ages*, Cambridge: Cambridge University Press.

Postan, M.M. (1972) *The Medieval Economy and Society: An Economic History of Britain in the Middle Ages*, Harmondsworth: Penguin.

Pound, J.F. (ed.) (1971) *The Norwich Census of the Poor, 1570*, Norwich: Norfolk Record Society.

Pound, J.F. (1988) *Tudor and Stuart Norwich*, Chichester: Phillimore.

Pred, A. (1977) 'The Choreography of Existence: Some Comments on Hägerstrand's Time-Geography and its Usefulness', *Economic Geography*, 53, 207–21.

Prior, L. (1993) *The Social Organization of Mental Illness*, London: Sage.

Rabinach, A. (1990) *The Human Motor: Energy, Fatigue, and the Origins of Modernity*, New York: Basic Books.

Radford, J. (1994) 'Intellectual Disability and the Heritage of Modernity', in Rioux, M.H. and Bach, M. (eds), *Disability is Not Measles: New Research Paradigms in Disability*, Ontario: Roeher Institute.

Radford, J.P. and Park, D.C. (1993) 'The Asylum as Place: an Historical Geography of the Huronia Research Centre', in Gibson, J.R. (ed.), *Canada: Geographical Interpretations, Essays in Honour of John Warkentin*, York University, Department of Geography, Monograph no. 22.

Rawls, J. (1971) *A Theory of Justice*, Cambridge, MA: Harvard University Press.

Rawls, J. (1993) *Political Liberalism*, New York: Columbia University Press.

Rea, D.M. (1995) 'Unhealthy Competition: the Making of a Market for Mental Health', *Policy and Politics*, 23, 2, 141–155.

Riley, J.C. (1987) 'Sickness in an Early Modern Workplace', *Continuity & Change*, 2, 3, 363–385.

Rioux, M.H. and Bach, M. (eds) (1994) *Disability is Not Measles: New Research Paradigms in Disability*, Ontario: Roeher Institute.

Ripper, P. (1997) 'Institutional Reform and Disability Legislation', *Abstract: The Newsletter of the Australian Disability Network*, 1, 1, 21.

Ronalds, C. (1990) *National Employment Initiatives for People with Disabilities – a Discussion Paper*, Canberra: Australian Government Publishing Service.

Roof, J. and Weigman, R. (eds) (1995) *Who Can Speak? Authority and Critical Identity*, Urbana: University of Illinois Press.

Rose, D. (1989) 'A Feminist Perspective of Employment Restructuring and Gentrification: the Case of Montreal', in Wolch, J. and Dear, M. (eds), *The Power of Geography: How Territory Shapes Social Life*, Boston: Unwin Hyman.

Rose, G. (1993) *Feminism and Geography: the Limits to Geographical Knowledge*, Cambridge: Polity.

Rosenthal, J.T. (1972) *The Purchase of Paradise: Gift Giving and the Aristocracy, 1307–1485*, London: Routledge & Kegan Paul.

Ross, K. (1988) *The Emergence of Social Space: Rimbaud and the Paris Commune*, Basingstoke: Macmillan.

Rothman, D.J. (1971) *The Discovery of the Asylum: Social Order and Disorder in the New Republic*, Boston: Little, Brown.

Rowe, S. and Wolch, J. (1990) 'Social Networks in Time and Space: Homeless Women in Skid Row, Los Angeles', *Annals of the Association of American Geographers*, 80, 2, 184–204.

Ruddick, S. (1997) *Young and Homeless in Hollywood: Mapping Social Identities*, New York: Routledge.

Ryan, J. and Thomas, F. (1987) *The Politics of Mental Handicap*, London: Free Association.

Safilios-Rothschild, C. (1970) *The Sociology and Social Psychology of Disability and Rehabilitation*, New York: University Press of America.

Sandel, M. (1982) *Liberalism and the Limits of Justice*, Cambridge: Cambridge University Press.

Savas, E.S. (1982) *Privatizing the Public Sector: How to Shrink Government*, Chatham, NJ: Chatham House.

Sayer, A. (1984) *Method in Social Science: A Realist Approach*, London: Hutchinson.

Scarry, E. (1985) *The Body in Pain: The Making and Unmaking of the World*, New York: Oxford University Press.

Scheer, J. and Groce, N. (1988) 'Impairment as a Human Constant: Cross-Cultural and Historical Perspectives on Variation', *Journal of Social Issues*, 44, 1, 23–37.

Scott, A.J. (1980) *The Urban Land Nexus and the State*, London: Pion.

Segalen, M. (1983) *Love and Power in the Peasant Family: Rural France in the Nineteenth Century*, Oxford: Blackwell.

Shakespeare, T. (1994) 'Cultural Representation of Disabled People: Dustbins for Disavowal?', *Disability and Society*, 9, 3, 283–299.

Shannon, P. (1991) *Social Policy*, Auckland: Oxford University Press.

Shannon, P.T. and Hovell, K.J., (1993) *Community Care Facilities: Experience and Effects*, Report prepared for Dunedin City Council and the Otago Area Health Board, Dunedin, New Zealand.

Shapiro, J.P. (1993) *No Pity: People with Disabilities Forging a New Civil Rights Movement*, New York: Times.

Sharpe, J.A. (1987) *Early Modern England: A Social History, 1550–1760*, London: Edward Arnold.

Sibley, D. (1995) *Geographies of Exclusion: Society and Difference in the West*, London: Routledge.

Slack, P. (ed.) (1975) *Poverty in Early Stuart Salisbury*, Devizes: Wiltshire Record Society.

Smith, C.J. (1978) 'Problems and Prospects for a Geography of Mental Health', *Antipode*, 10, 1, 1–12.

Smith, C.J (1981) 'Urban Structure and the Development of Natural Support Systems for Service Dependent Populations', *The Professional Geographer*, 33, 457–465.

Smith, C.J. (1984) 'Geographical Approaches to Mental Health', in Freeman, H. (ed.), *Mental Health and the Environment*. London: Churchill Livingstone.

Smith, C.J. (1989) 'Privatisation and the delivery of mental health services', *Urban Geography*, 6, 88–99.

Smith, C.J. and Giggs, J.A. (1988) 'Introduction', in Smith, C.J. and Giggs, J.A. (eds), *Location and Stigma: Contemporary Perspectives on Mental Health and Mental Health Care*, Boston: Unwin Hyman.

Smith, D.M. (1977) *Human Geography: a Welfare Approach*, London: Edward Arnold.

Smith, D.M. (1994) *Geography and Social Justice*, Oxford: Blackwell.

Smith, N. (1979) 'Geography, Science and Post-Positivist Modes of Explanation', *Progress in Human Geography*, 3, 356–383.

Smith, N. (1984) *Uneven Development*, Oxford: Blackwell.

Smith, N.J. and Smith, H.C. (1991) *Physical Disability and Handicap: A Social Work Approach*, Melbourne: Longman Cheshire.

Smith, R., Gaster, L., Harrison, L., Martin, L., Means, R. and Thistlethwaite, P. (1993) 'Introduction', in Smith, R., Gaster, L., Harrison, L., Martin, L., Means, R. and Thistlethwaite, P. (eds), *Working Together for Better Community Care*, Bristol: School of Advanced Urban Studies.

Smith, S.R. and Lipsky, M. (1993) *Nonprofits for Hire: the Welfare State in the Age of Contracting*, Cambridge, MA: Harvard University Press.

Smull, M.W. (1990) 'Crisis in the Community', *Interaction*, 4, 3, 25–39.

Soja, E. (1989) *Postmodern Geographies: the Reassertion of Space in Social Theory*, London: Verso.

Soper, K. (1979) 'Marxism, Materialism and Biology', in Mepham, J. and Ruben, D-H. (eds), *Issues in Marxist Philosophy: Volume Two – Materialism*, Brighton: Harvester.

Soper, K. (1981) *On Human Needs: Open and Closed Theories in a Marxist Perspective*, Brighton: Harvester.

Soper, K. (1995) *What is Nature?*, Oxford: Blackwell.

Stallybrass, P. and White, A. (1986) *The Politics and Poetics of Transgression*, London: Methuen.

Steinman, L.D. (1987) 'The Effect of Land-Use Restrictions on the Establishment of Community Residences for the Disabled: a National Study', *The Urban Lawyer*, 19, 1–37.

Stewart, B. (1993) 'New Human Rights Bill is a Big Advance', *New Zealand Disabled*, July, 8–10.

Stone, D. (1984) *The Disabled State*, Philadelphia: Temple University Press.

Stuart, O. (1992) 'Race and Disability: What Type of Double Disadvantage?', *Disability, Handicap and Society*, 7, 2, 177–188.

Stuart, O. (1993) 'Double Oppression: an Appropriate Starting Point?', in Swain, J., Finkelstein, V., French, S. and Oliver, M. (eds) *Disabling Barriers – Enabling Environments*, London: Sage, 93–100.

Sturrock, J. (1986) *Structuralism*, London: Paladin.

Swain, S. (1985) 'The Poor of Melbourne', in Davison, G., Dunstan, D. and McConville, C. (eds), *The Outcasts of Melbourne: Essays in Social History*, Sydney: Allen & Unwin, 91–112.

Swain, J., Finkelstein, V., French, S. and Oliver, M. (1993) 'Introduction' in Swain, J., Finkelstein, V., French, S. and Oliver, M. (eds) *Disabling Barriers – Enabling Environments*, London: Sage, 1–7.

Swerdlow, J.L. (1995) 'Information Revolution', *National Geographic*, October, 5–29.

Taylor, S.M. (1988) 'Community Reactions to Deinstitutionalization', in Smith, C.J. and Giggs, J.A. (eds), *Location and Stigma: Contemporary Perspectives on Mental Health and Mental Health Care*, London: Unwin Hyman.

Thomas, H. (1992) 'Disability, Politics and the Built Environment', *Planning Practice and Research*, 7, 1, 22–24.

Thompson, E.P. (1974) 'Time, Work-Discipline, and Industrial Capitalism', in Flinn, M.W. and Smout, T.C. (eds), *Essays in Social History*, Oxford: Clarendon Press.

Thompson, G. (1990) *The Political Economy of the New Right*, London: Pinter.

Thomson, R.G. (1997) *Extraordinary Bodies: Figuring Physical Disability in American Culture and Literature*, Irvington, NY: Columbia University Press.

Thorpe, C. and Toikka, R. (1980) 'Determinants of Racial Differentials in Social Security Benefits', *Review of Black Political Economy*, 10, 4.

Thrift, N. (1981) 'Owners' Time and Own Time: The Making of a Capitalist Time Consciousness, 1300–1800', in Pred, A. (ed.), *Space and Time in Geography: Essays Dedicated to Torsten Hagerstrand*, WK Gleerup.

Thrift, N. (1990) 'The Making of a Capitalist Time Consciousness', in Hassard, J. (ed.), *The Sociology of Time*, London: Macmillan.

Timpanaro, S. (1975) *On Materialism*, London: New Left Books.

Tisato, P. (1997) 'Travel Affordability for People with Disabilities', *Urban Policy and Research*, 15, 3, 175–187.

Topliss, E. (1982) *Social Responses to Handicap*, London: Longman.

Townsend, P. (1979) *Poverty in the United Kingdom*, Harmondsworth: Penguin.

Tronto, J.C. (1987) 'Beyond Gender Difference to a Theory of Care', *Signs*, 12, 4, 644–663.

Tronto, J.C. (1993) *Moral Boundaries: A Political Argument for an Ethic of Care*, New York: Routledge.

Turner, B.S. (1984) *The Body and Society: Explanations in Social Theory*, Oxford: Blackwell.

Turner, B.S. (1991) 'Recent Developments in the Theory of the Body', in Featherstone, M., Hepworth, M. and Turner, B.S. (eds), *The Body, Social Process and Cultural Theory*, London: Sage.

Union of Physically Impaired Against Segregation (UPIAS) (1976) *Fundamental Principles of Disability*, London: UPIAS.

Ure, A. (1967 [1835]) *The Philosophy of Manufactures or an Exposition of the Scientific, Moral and Commercial Economy of the Factory System*, London: Frank Cass.

Urmson, J.O. and Ree, J. (eds) (1991) *The Concise Encyclopedia of Western Philosophy and Philosophers*, London: Unwin Hyman.

Vladeck, B.C. (1980) *Unloving Care: the Nursing Home Tragedy*, New York: Basic Books.

Vogel, L. (1983) *Marxism and the Oppression of Women: Toward a Unitary Theory*, London: Pluto Press.

Vujakovic, P. and Matthews, M.H. (1992) *Mapping Another World: Physical Disabilities and the Urban Environment*. Cambridge: Anglia Polytechnic, Division of Geography, GEOinformatics Unit Handbook and Report Series No. 1.

Vujakovic, P. and Matthews, M.H. (1994) 'Contorted, Folded, Torn: Environmental Values, Cartographic Representation and the Politics of Disability', *Disability and Society*, 9, 3, 359–374.

Walker, A. (1980) 'The Social Creation of Poverty and Dependency in Old Age', *Journal of Social Policy*, 9, 1, 49–75.

Walker, A. (1989) 'Community Care', in McCarthy, M. (ed.), *The New Politics of Welfare*, London: Macmillan.

Walker, R. (1981) 'A Theory of Suburbanisation: Capitalism and the Construction of Urban Space in the United States', in Dear, M. and Scott, A.J. (eds), *Urbanization and Urban Planning in Capitalist Societies*, New York: Methuen.

Wallerstein, I. (1983) *Historical Capitalism*, London: Verso.

Walzer, M. (1983) *Spheres of Justice, A Defence of Pluralism and Equality*, New York: Basic Books.

Warren, B. (1980) 'Some Thoughts Towards a Philosophy of Physical Handicap', in Laura, R.S. (ed.), *The Problem of Handicap*, Melbourne: Macmillan.

Watson, F. (1930) *Civilisation and the Cripple*, London: John Bale, Sons and Danielsson.

Welch, R.V. (1996) 'Dunedin', in Le Heron, R. and Pawson, E. (eds), *Changing Places: New Zealand in the Nineties*, Auckland: Longman.

Wendell, S. (1989) 'Towards a Feminist Theory of Disability', *Hypatia*, 4, 2, 104–124.

Wendell, S. (1996) *The Rejected Body*, London: Routledge.

Wibberly, G.P. (1978) 'Mobility and the Countryside' in Cresswell, R. (ed.), *Rural Transport and Country Planning*, London: Leonard Hill.

Williams, C. and Thorpe, B. (1982) *Beyond Industrial Sociology: the Work of Women and Men*, Sydney: Allen & Unwin.

Williams, R. (1978) 'Problems of Materialism', *New Left Review*, 109, 3–17.

Williams, R. (1980) *Problems in Materialism and Culture – Selected Essays*, London: Verso.

Wilmot, S. (1997) *The Ethics of Community Care*, London: Cassell.

Winzer, M.A. (1997) 'Disability and Society before the Eighteenth Century', in Davis, L.J. (ed.) *The Disability Studies Reader*, New York: Routledge.

Wionarski, G. and Abbott, E.S. (1945) *Women Who Helped Pioneers: Pages of Melbourne's History that Glow*, Melbourne: Melbourne Ladies' Benevolent Society.

Wohl, A.S. (1983) *Endangered Lives: Public Health in Victorian Britain*, London: Methuen.

Wolch, J. (1980) 'The Residential Location of the Service-Dependent Poor', *Annals of the Association of American Geographers*, 70, 330–341.

Wolch, J. (1989) 'The Shadow State: Transformations in the Voluntary Sector', in Wolch, J. and Dear, M. (eds), *The Power of Geography: How Territory Shapes Social Life*, Boston: Unwin Hyman, 197–221.

Wolch, J. (1990) *The Shadow State: Government and the Voluntary Sector in Transition*, New York: The Foundation Center.

Wolch, J. and Dear, N. (1993) *Malign Neglect: Homelessness in an American City*, San Francisco: Jossey-Bass.

Wolfensberger, W. (1983) 'Social Role Valorization: a Proposed New Term for the Principle of Normalization', *Mental Retardation*, 21, 6, 234–239.

Wolfensberger, W. (1987) 'The Ideal Human Service', *Interaction – The Australian Magazine on Intellectual Disability*, 2, 3–4.

Wolfensberger, W. (1995) 'Social Role Valorization is too Conservative. No it is too Radical', *Disability and Society*, 10, 3, 365–367.

Wolfensberger, W. and Nirje, B. (1972) *The Principle of Normalization in Human Services*, Toronto: National Institute on Mental Retardation.

Wolfensberger, W. and Thomas, S. (1983) *Passing: Program Analysis of Service Systems' Implementation of Normalization Goals*, 2nd edn, Toronto: National Institute on Mental Retardation.

Wolpert, E. and Wolpert, J. (1974) 'From Asylum to Ghetto', *Antipode*, 6, 63–76.

Wolpert, J. (1976) 'Opening Closed Spaces', *Annals of the Association of American Geographers*, 66, 1, 1–13.

Wolpert, J. (1978) *Group Homes for the Mentally Retarded and Investigation of Neighbouring Property Impacts*, Princeton, NJ: Woodrow Wilson School of Public and International Affairs, Princeton University.

Wolpert, J. (1980) 'The Dignity of Risk', *Transactions, Institute of British Geographers*, 5, 4, 391–410.

Wood, A. (1981) *Karl Marx*, London: Routledge & Kegan Paul.

Wood, R. (1991) 'Care of Disabled People' in Dalley, G. (ed.), *Disability and Social Policy*, London: Policy Studies Institute.

Wrightson, W. (1989) *From Barrier Free to Safe Environments: the New Zealand Experience*, New York: World Rehabilitation Fund.

Yeatman, A. (1996) *Getting Real: the Interim Report of the Review of the Commonwealth/State Disability Agreement*, Canberra: Australian Government Publishing Service.

Young, I.M. (1990) *Justice and the Politics of Difference*, Princeton NJ: Princeton University Press.

Young, I.M. (1997) 'Unruly Categories: a Critique of Nancy Fraser's Dual Systems Theory', *New Left Review*, 222, March/April, 147–160.

Zipple, A. and Anzer, T.C. (1994) 'Building Code Enforcement: New Obstacles in Siting Community Residences', *Psychosocial Rehabilitation Journal*, 18, 1, 5–13.

Zola, I. (1993) 'Self, Identity and the Naming Question: Reflections on the Language of Disability', *Social Science and Medicine*, 36, 2, 167–173.

Index

Abberley, P. 5, 9, 15–25 *passim*, 27
abbeys 93, 100
abject bodies 110–111, 119, 121, 136
'ableism' 27, 130, 134
abnormality 17, 135
abuse: entrenched 'culture of' 155;
    sexual 131, 155, 166
access(ibility) 16; state social services
    28; transit services 165; urban
    173–194; *see also* inaccessibility
accountability 165
activism 140–141
ADAPT (American Disabled for
    Accessible Public Transport) 141,
    166–167
affective relations 146, 147, 148
affirmative 'remedies'/action 144–145,
    178
agriculture 77, 79, 87, 101, 102
AIDS 202
Alcock, P. 134, 138
allocation of resources 145
almshouses 93, 94, 95
Althusser, L. 45
anatomy 41
Anderson, P. 75, 76, 77
Anglophilia 112
Annales school 66, 67–68, 75
anthropology 65
Antiquity: Classical 25, 55; Islamic 24
Anzer, T.C. 158
'appearance' 26
approbation 22
aptitude 54
Archer, Thomas 110
architecture 28, 29, 137
aristocracy 75
asylums 108, 114–115; 'duty to attend'
    110

attitudes 25, 105, 186; changing 22,
    26; discriminatory 185; negative 21;
    patronising 185; socio-cultural 154
Auckland 191
Ault, W.O. 86
Australia 4, 133, 134, 135, 139, 140,
    173; activism 141; community care
    153, 155, 158, 161, 164, 167, 168;
    Disability Discrimination Act (1992)
    143; disability social movements 143;
    Human and Equal Opportunities
    Commission 141; multi-provider
    model 153; NIMBY reactions 158;
    see also Melbourne; Sydney
'avoidance strategies' 158

Badcock, B. 145
Barnes, C. 24, 26, 27, 133, 155
Barrett, B. 112
Barton, Len 15
beauty 54
beggars 92, 103, 118; 'crippled' 110,
    124; 'hideous' 124; lame 61, 62, 63;
    women 123, 124
behaviouralist perspective 28, 29
Beier, A.L. 92
beliefs 64
'benevolence' 108, 114, 116
Bennie, G. 168
Benthall, J. 43
Berens, B.S. 24
Berthoud, R. 134
Bewley, C. 164
Bickenbach, J. 20
biology 37, 40, 41, 44
'bivalence' 131, 134, 147
Black Death (1348–49) 103
black people 17
blind people 30–31